Landscapes of Mars
A VISUAL TOUR

Landscapes of
Mars
A VISUAL TOUR

GREGORY L. VOGT

Springer

© 2008 Springer Science+Business Media, LLC

Printed on acid-free paper.

Library of Congress Control Number: 2008933361

Printed in the United States of America.

9 8 7 6 5 4 3 2 1

springer.com

Landscapes of Mars
A VISUAL TOUR

CONTENTS

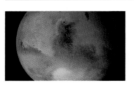
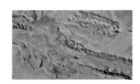
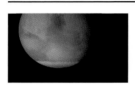
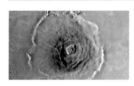

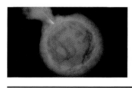

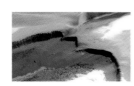
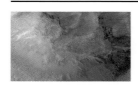

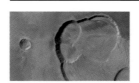

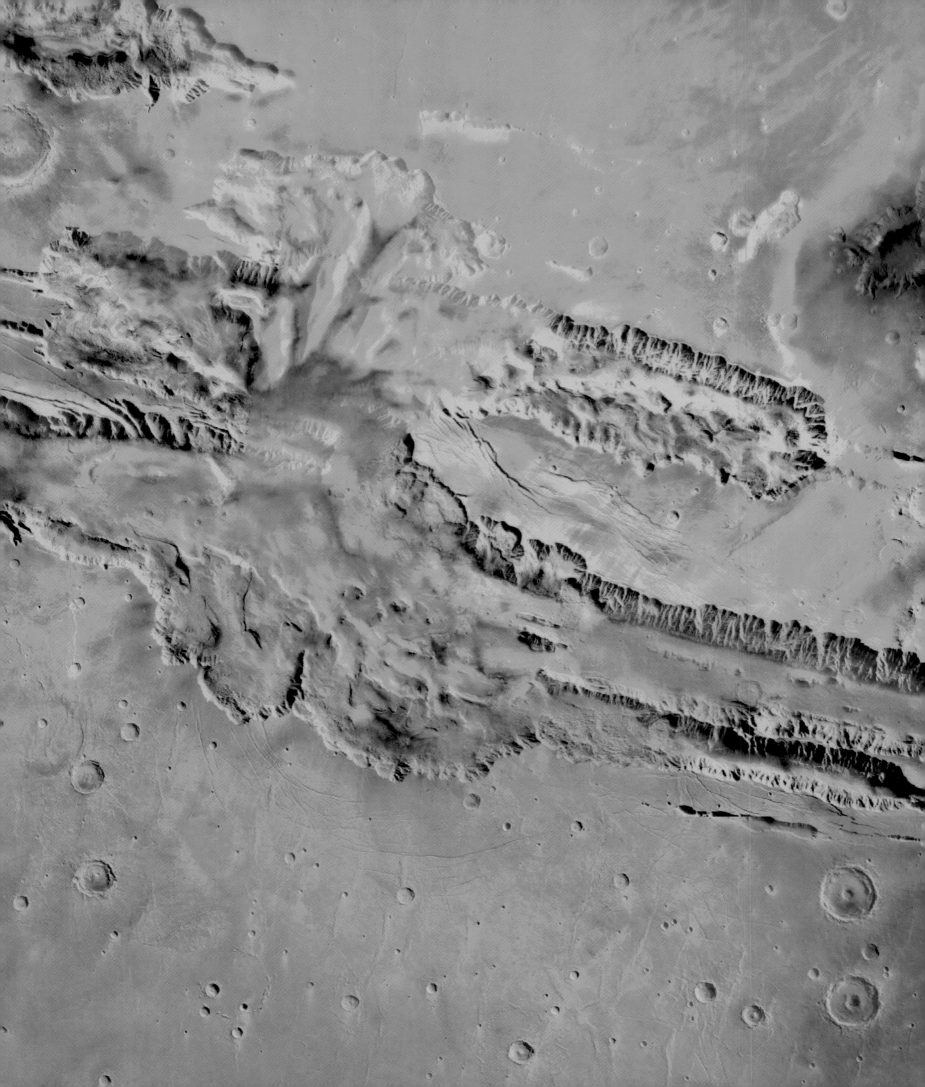

INTRODUCTION

Practically next door to us in our Solar System, Mars is the most Earth-like of all the worlds held in perpetual motion by the Sun's gravity. It is a planet that has fascinated and intrigued humans since prehistoric times. The Greeks named it Ares, after their god of war. The connection was easy to make because of the planet's almost blood-red color. The Romans were also impressed by the color and renamed the planet Mars, after their war god.

Mars's bloody color was not the only thing that captured the attention of the ancients. The planet is the fourth brightest nighttime object after the Moon, Venus, and Jupiter, but its brightness is not steady. Sometimes Mars is faint, like the surrounding stars, while at other times it is a beacon. Furthermore, Mars's motion appears to be chaotic, suggesting the state of disorder common during times of war.

Mars slowly travels eastward, changing its position among the background stars. This movement granted Mars planet status, along with Mercury, Venus, Jupiter, and Saturn. The word planet comes from the Greek word planetes, meaning "wanderer."

This is Valles Marineris, the great canyon of Mars, as seen by Viking. The composite scene shows the entire canyon system, over 3,000 km long and averaging 8 km deep. The connected chasma, or valleys, of Valles Marineris may have formed from a combination of erosional collapse and structural activity. Layers of material in the eastern canyons might consist of carbonates deposited in ancient lakes. Huge ancient river channels ran north from Valles Marineris and from adjacent canyons. Many of the channels flowed into Chryse Basin, which contains the site of the Viking 1 and Mars Pathfinder landers.
Image Credit: NASA/JPL/USGS

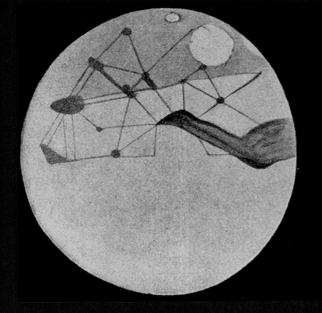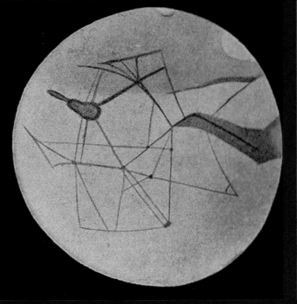

Mars "channels" as sketched as observed by Percival Lowell sometime before 1914. South, here, is at the top.

Alien tripod illustration from an early French edition of H. G. Wells' The War of the Worlds.

The motion of Mars drew special attention because it would slow in its eastward course and loop backward for a time before returning to its original path. The other classical planets also exhibited the property of backwards, or retrograde, motion. However, because of vastly greater or shorter distances to the Sun, these motions were less noticeable. We know today that Mars's perplexing movements and brightness changes are the result of the orbital dance between Earth and Mars. However, in ancient times, the mysterious phenomenon led to myth and foretelling.

Much later, as telescopes became the instrument of choice for exploring the night sky, the Martian surface came into view. Although Mars resolved into a disk, unlike background stars, which remained pinpoints, the surface appeared fuzzy. Observer sketches revealed a surface marked by shadowy features that changed over time. As instrumentation improved, the shadows became more detailed, enabling observers to determine the tilt of Mars's axis and the length of its day. Light-colored polar caps were discovered that grew and shrank. Except for an orbit twice as long and a diameter half the size of Earth's, Mars is beginning to look like Earth's younger sibling. The shadow variations could easily be explained. Springtime polar cap melting yielded water for exploding summer plant growth. In fall, the plants would go to seed and desiccate as water is refrozen into the polar caps in colder weather.

In the minds of some observers, the changeable Martian shadows morphed into regular patterns, a latticework of channels that could indicate intelligence at work. If purposely created, the channels would actually be canals designed to distribute polar meltwater to the drier equatorial regions.

For a time, Martian canal fever was infectious. With each new observation, hand-drawn surface maps revealed increasingly complex networks of intersecting canals, parallel canals, and oases at intersection points. In popular fiction, most notably the 1897 Pearson Magazine serialization of H.G. Well's "War of the Worlds," Mars was populated by a superior race of alien beings that looked upon Earth as a destination, a treasure house of resources and food (human blood) waiting to be taken.

The channel/canal "discoveries" triggered a multi-generational scientific debate about the true nature of the Martian surface. The debate ended in the1970s, however, when Earth spacecraft orbited the Red Planet and began detailed mapping of its surface. Martian canals devolved into individual features and surface color variations, which linked together as linear traces in the minds of hopeful observers. What the spacecraft found instead of canals was equally stunning.

Map of Mars by Giovanni Schiaparelli

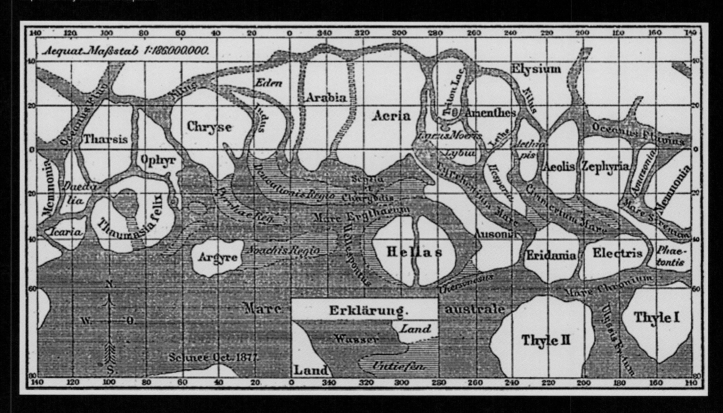

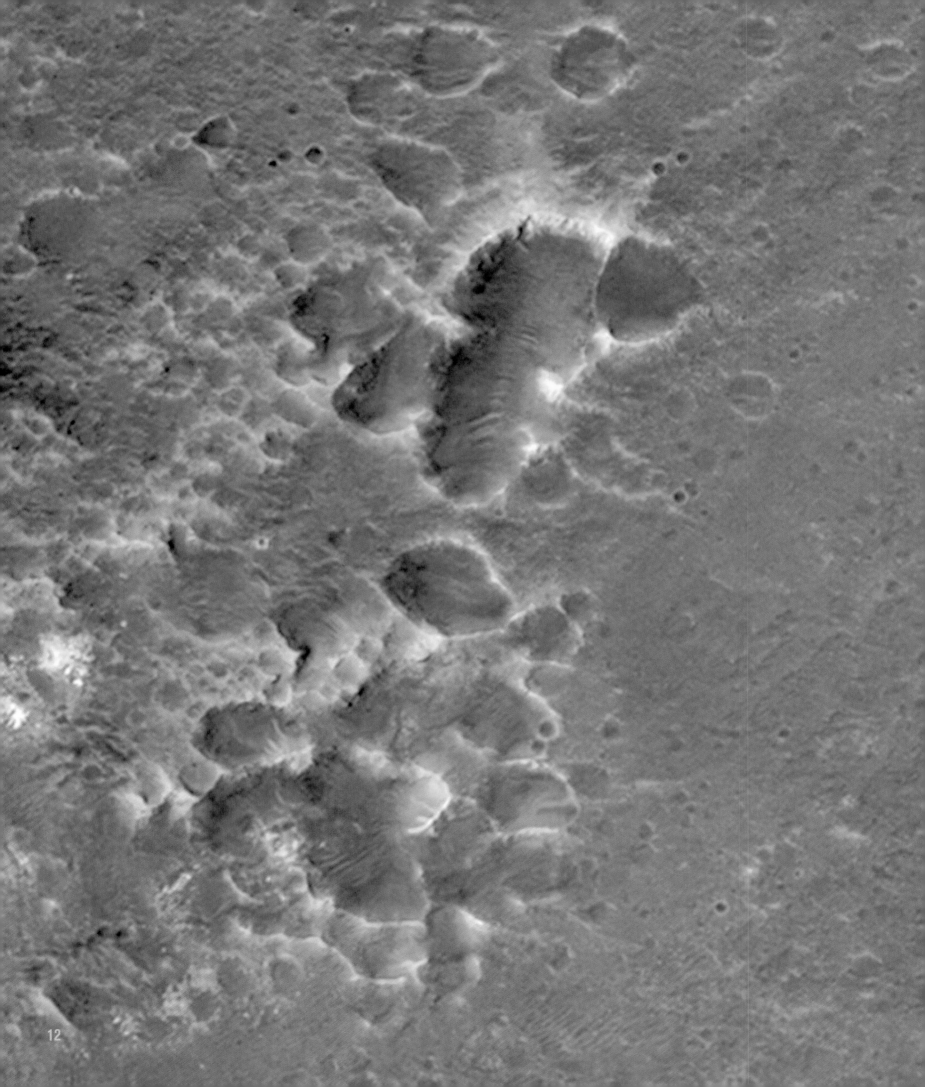

Mawrth Vallis shows a rich mineral diversity, including clay minerals that formed by the chemical alteration of rocks or loose "regolith" (soil) by water. The CRISM instrument on the MRO spacecraft has detected a variety of those minerals here, which could signify different processes of formation. This surface is scientifically compelling for the Mars Science Laboratory (MSL) rover, although some of the terrain can be somewhat rough.
Image credit:
NASA/JPL/University of Arizona

Mariner 4 was the first successful Mars space probe. After seven and a half months in flight, Mariner 4 passed Mars at a distance of just under 10,000 kilometers. Its science mission lasted two days. Mariner 4's camera took 21 pictures plus a partial image of picture 22 on July 14 and 15, 1965. The pictures totaled were grainy and covered a swath of ancient meteor-pocked terrain. Other onboard instruments recorded very low atmospheric pressures and temperatures of -100 degrees Celsius.
Image credit: National Space Science Data Center, NASA Goddard Space Flight Center

Today, in a reverse on science fiction, Mars is capturing serious attention as a possible destination for human expansion. Although severely lacking in atmosphere and offering shatteringly low temperatures and deadly ionizing radiation, future human colonies on Mars are possible. In spite of its shortcomings, Mars is the most Earth-like of the known planets. Dreamers in space agencies around the world are no longer talking about sending humans to Mars someday. They are actively investigating the Martian environment and surface topography, making sure there will be no major surprises when astronauts arrive. They are lobbying for resources and testing the technology to make Martian colonies self-sustaining. They are even laying the plans for the transport rockets, surface vehicles, habitats, space suits, and power systems. The first steps of going to Mars are being taken.

In the history of human settlement of Earth, mostly small groups of explorers have set out looking for new worlds and resources. Explorers were followed by pioneers and, eventually, by civilization. In our move toward Mars, we have sent our first explorers. In 1960, the first interplanetary robotic probe was launched by the Soviet Union. Marsnik 1 failed during launch. It was followed over the next two years by four other Soviet Mars probes. Three experienced launch failures, but Mars 1 provided valuable micrometeorite, magnetic field, and solar wind data before contact was lost with Mars still 87 million km (53,940,000 m) away. Mariner 4, a U.S. space probe, attempted the interplanetary trip in 1964. After a 7-month voyage, Mariner 4 arced around Mars, passing about 9,850 km above the surface (about 6,000 m). Its cameras took 21 close-up pictures. Mariner 4's pictures and instrument data succeeded in repainting the Martian surface. No signs of life or canals or oases were found, but the pictures did reveal lots of craters. Sensors measured daytime temperatures of around -100° C (-148° F) and detected a very thin atmosphere. Mars appeared to be more Moon-like than Earth-like.

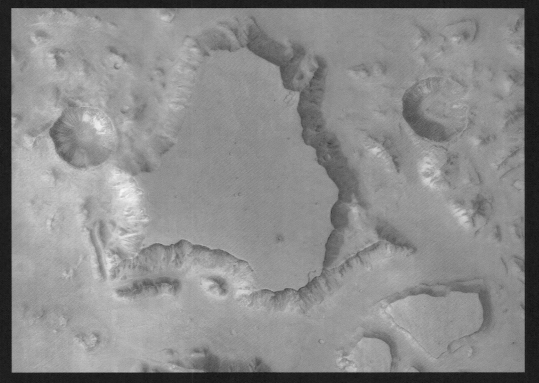

This picture was taken by the High Resolution Stereo Camera (HRSC) onboard ESA's Mars Express orbiter in January 2004. It shows a vertical view of a mesa in the true colors of Mars. The summit plateau stands about 3 km above the surrounding terrain. The original surface was dissected by erosion, and only isolated mesas remained intact. The large crater has a diameter of 7.6 km
Image credit: ESA/DLR/FU Berlin
(G. Neukum)

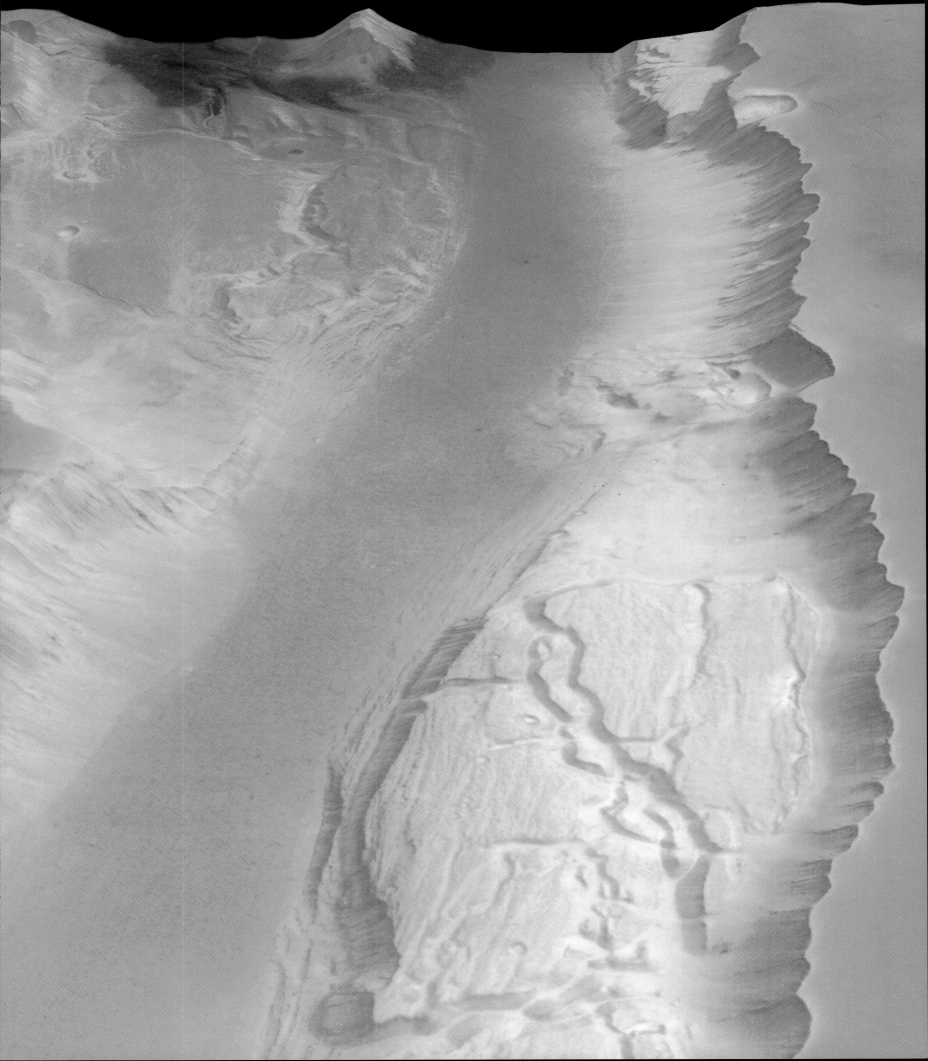

The rover Opportunity in 2006 captured this vista of "Victoria Crater" from the viewpoint of "Cape Verde," one of the promontories that are part of the scalloped rim of the crater. The view combines hundreds of exposures taken by the rover's panoramic camera (Pancam).
Image credit: NASA/JPL-Caltech/Cornell

Fortunately, Mariner 4 was only the beginning of direct Martian exploration. Mariner 4's photographic close-ups represented only 1 percent of the total planetary surface. (By coincidence, the entire surface area of Mars is approximately equal to the total dry land surface of Earth.) It would be like judging all of Earth by looking just at the state of Alaska. As it turned out, Mariner 4 captured images of some of the least interesting areas of Mars. These pictures prompted some scientists to declare that Mars was a geologically dead world, words they were to regret some eight years later.

In 1971, a different kind of Mariner spacecraft, Mariner 9, reached Mars and went into orbit. In the worst of coincidences, astronomers were startled by a planet-wide dust storm that began forming two months before Mariner 9 was scheduled to slip into orbit on November 14. No surface features were visible except the summits of Olympus Mons and the other Tharsis volcanoes. The storm began abating in late November, and by the time Mariner 9 ceased operations, it had collected over 7,300 images covering the entire planet. Mars had a new face: a world of broad plains, mountains, impact craters, channels, canyons, and volcanoes dwarfing the largest similar features on Earth. The Mariner spacecraft had revealed a frigid desert world of the familiar and the fantastic.

Robot explorers continue to probe the Red Planet but with vastly improved instruments and sensors. There is now a multinational armada of Mars spacecraft in orbit, roving the surface, or still crunched in bits and bytes in the hard drives of modern "drawing boards." It will soon be time for the first human explorers, and then will come the pioneers. What will they see and experience? The time for Mars has come. The chapters that follow take you on a visual tour of the landscapes of the next New World.

More than 10 percent of the far northern surface area on Mars is covered by windblown sand dunes. This HiRISE image shows the location where in 2008 the first significant change to sand dunes was reported on Mars. That study used a time series of MOC images taken over a period of three Martian years and showed that two 20-meter-wide dome dunes disappeared and a third shrank by an estimated 15 percent. Other, larger dunes in the same location do not show apparent change: more time or more precise measurements to display evidence of change is needed.
Image Credit: Credit: NASA/JPL/University of Arizona

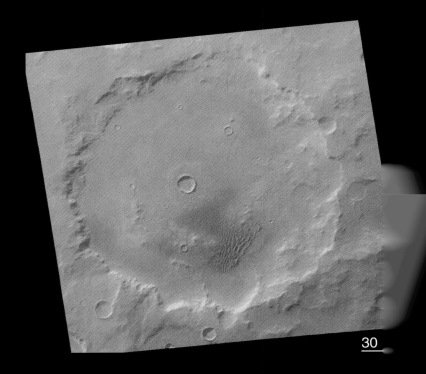

As the MGS primary mission draws to an end, the southern hemisphere of Mars is in the depths of winter. At high latitudes, it is dark most, if not all, of the day. Even at middle latitudes, the sun shines only thinly through a veil of water and carbon dioxide ice clouds, and the ground is so cold that carbon dioxide frosts have formed. Kaiser Crater is one such place. At a latitude comparable to Seattle, Washington, Kaiser Crater is studied primarily because of the sand dune field found within the confines of its walls (lower center of the MOC image, above). Close-up pictures of these and other dunes in the region show details of their snow cover, including small avalanches.
Image credit: NASA/JPL/MSSS

NASA's Phoenix Mars lander monitors the atmosphere overhead and reaches out to the soil below in this artist's depiction of the spacecraft fully deployed on the surface of Mars. Phoenix landed in the spring of 2008 on an arctic plain of northern Mars, where there are indications of frozen water mixed with soil within arm's reach of the surface. It will use its robotic arm to dig down to the expected icy layer and analyze scooped-up samples for factors that will help scientists evaluate whether the subsurface environment at the site was or still is a favorable habitat for microbial life. A weather station (see vertical green line in illustration) on the lander will conduct the first study Martian arctic weather from ground level. The dark "wings" to either side of the lander's main body are solar panels used for providing electric power.
Image Credit: NASA/JPL/UA/Lockheed Martin

NASA's Mars Science Laboratory, a mobile robot for investigating Mars' past or present ability to sustain microbial life, is scheduled for launch in 2009. This artist's concept of the robot portrays what the advanced rover would look like in Martian terrain. The arm extending from the front of the rover is designed both to position some of the rover's instruments onto selected rocks or soil targets and also to collect samples for analysis by other instruments. The mast, rising to about 2.1 meters (6.9 feet) above ground level, supports two remote-sensing instruments for viewing of surrounding terrain and analyzing the types of atoms in material that laser pulses have vaporized from rocks or soil targets up to about 9 meters (30 feet) away.

Image credit: NASA/JPL-Caltech

1 THE TWO FACES OF MARS

Whether or not Mars is Earth-like, it is a unique world in a Solar System populated by distinctly unique planets. Mars is classified as a terrestrial or rocky planet, along with Mercury, Venus, and Earth. Beyond Mars and beyond the wide and loosely packed Asteroid Belt are the giant gas planets. Jupiter, the largest of all, contains more matter than all the Sun's other planets, asteroids, and comets combined. On a size scale, Mars is the second smallest planet after Mercury (or third smallest if you are still holding out for Pluto). It is only one half the diameter of Earth.

A planet's physical characteristics are dependent upon a wide range of environmental factors, including size, distance from the Sun, rotation rate, axial tilting, composition, and atmospheric density. Starting with Mercury, we find a world that is thoroughly baked by the Sun. It rotates one and a half times with each 88-day orbit. With virtually no atmosphere to balance temperatures, Mercury's "weather" is either beastly hot (467° C, or 872° F) or beastly cold (-183° C, or -297° F), depending on which side you are on. Its surface is heavily cratered and moonlike.

These two spacecraft images show the size relationship between Earth and Mars. In spite of the obvious color differences, both planetary bodies contain water, though much of Mars's water is now locked up as ice beneath its surface. Mars is slightly larger than one half the diameter of Earth. Squished together, it would take about six and a half planets the size of Mars to equal Earth in volume. The surface area of Mars is approximately equal to the total exposed land surface of Earth.

Twice as far from the Sun, Venus is a volcanic world dotted with tens of thousands of volcanoes and lava-covered plains. It is Earth's twin in size. Planet-wide, Venus is as hot as the Sun-lit side of Mercury. It has an atmosphere with a surface pressure 90 times greater than Earth's. Consisting predominantly of carbon dioxide and perpetually cloudy, the atmosphere captures and holds the Sun's heat. Venus is a planet where global warming has gone wild.

Earth, third outward from the Sun, is sometimes referred to as the "Goldilocks Planet" because it is at the perfect distance from the Sun (neither too hot nor too cold). Earth is massive enough to maintain a substantial atmosphere, with a good balance of oxygen and nitrogen. Its temperature range is perfect for water to exist in its three forms—liquid, solid, and gas—and for many chemical reactions to take place. This combination makes Earth the garden planet of the Solar System and home to the only known life in the universe.

About half again further from the Sun than Earth is the orbital space that Mars inhabits. A small rocky body with a very thin atmosphere, Mars is a world of contrasts. Its surface contains a record of geological events that span 4 billion years. Today, Mars is home to some of the most dramatic geological formations in the Solar System. It also can be said that Mars has a split personality. Approximately half of the planet, the southern highlands, is mountainous. This area exhibits extensive cratering and much erosion and resembles the lunar highlands in appearance. Geologically, the highlands contain the most ancient surfaces of Mars.

The other half of Mars's surface encompasses the northern lowland areas. These are a broad and relatively smooth series of interconnecting plains comparable to the lunar lowlands. Their surface consists of basaltic lava flows and sedimentary deposits. (Weathering of the basalt leads to the rusty color of the planet.) These two faces of Mars are nearly equal hemispheres that are tilted to the planet's equator and separated in some places by a 2-km (1.2 m)-high scarp.

Although exaggerated in these images, the Martian topography is stunning. The planet is dotted with mammoth volcanic mountains, sliced by deep chasms, and divided by ancient cratered domains and broad, smooth plains. The darkest colors on these computer-generated maps represent the lowest elevations, while yellows, reds, and whites reveal the highest elevations. These two images were based on data collected by the Mars Orbiter Laser Altimeter carried by the Mars Global Surveyor Orbiter. The upper image depicts the Martian southern hemisphere while the lower depicts the northern hemisphere.

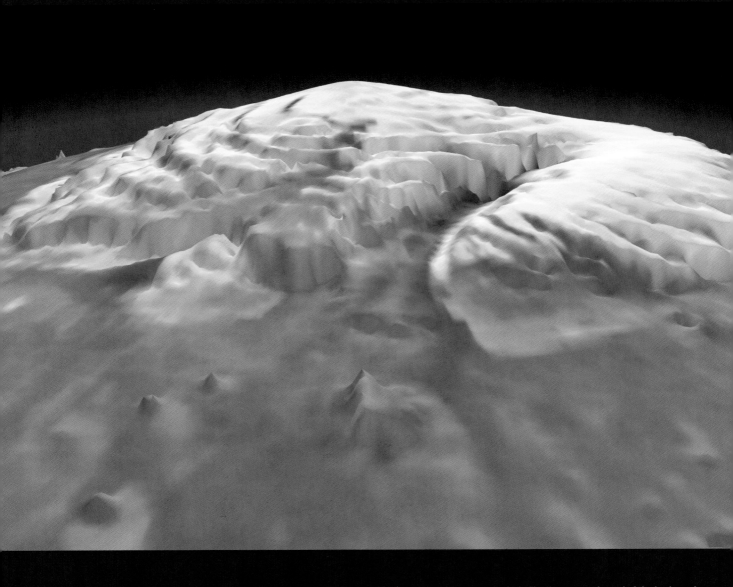

This first three-dimensional picture of Mars's north pole is not an anaglyph, like the ones at the end of the book, so do not use your glasses on it. However, the technology enabled scientists to estimate the volume of its water ice cap with unprecedented precision, and to study its surface variations and the heights of clouds in the region for the first time.

Capping Mars's rotational poles are large fields of water and dry ice that vary in surface extent and thickness with the seasons. Because of the rapid changes occurring at the poles, the polar regions exhibit some of the youngest surface features on Mars.

Mars orbits the Sun at a mean distance of 228 million km (142 million m). The orbit is fairly eccentric (elongated), causing it to vary from 207 million km (129 million m) at perihelion to 249 million km (155 million m) at aphelion. It takes Mars 687 days to orbit the Sun. Other orbital properties are Earth-like. Its day is 24 hours and 37 minutes long. It currently has a season-causing axial tilt of 25.2° (Earth's is 23.5°).

This false-color image shows the layered deposits on top of the north pole and the darker materials at the bottom, exposed in a scarp at the head of Chasma Boreale, a large canyon that eroded to form these layered deposits.

The deposits appear red because of dust mixed within them, but they are ice-rich, as indicated by previous observations. Water ice in the layered deposits is probably responsible for the pattern of fractures seen near the top of the scarp. The darker material below the layered deposits may have been deposited as sand dunes. It appears that brighter, ice-rich layers were deposited between the dark dunes in places. Exposures such as these are useful in understanding recent climate variations.

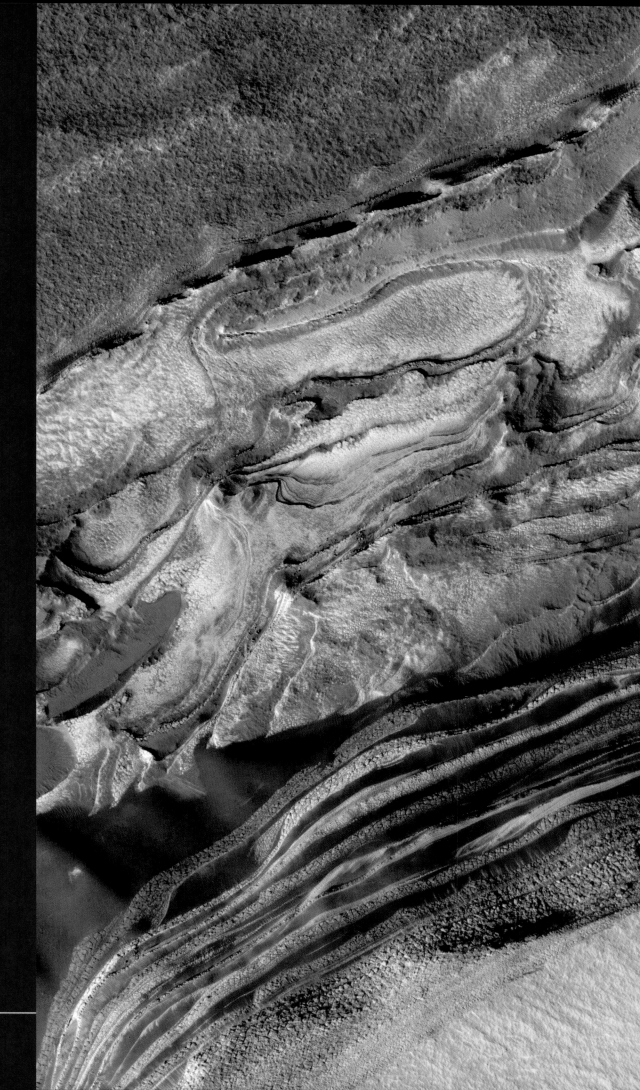

Martian Time

NOACHIAN ERA
4.5 to 3.5 billion years ago (+ or −
100 million years)

The Noachian Era began with the aggregation of solar nebula matter into the planets of the Solar System. It was a period of intensive volcanism and debris bombardment, energetic erosion, and cooling. The early Martian atmosphere was significantly denser than today, and watery bodies, perhaps oceans, covered large portions of the surface as the era was ending. Water flowed freely and sculpted the surface

HESPERIAN ERA
3.5 to 2.5 or 2.0 billion years ago
(+ or − 500 million years)

The Hesperian Era was a time of transition from the violence of the Noachian Era to the Mars of today. Mars experienced significant cooling, and surface water settled in and froze in subsurface rocks. Rivers continued flowing but at greatly reduced rates. Rapid melting of frozen groundwater led to quick "breakouts," resulting in massive local floods.

AMAZONIAN ERA
2.5 or 2.0 billion years ago to
Present (+ or − 500 million years)

The Amazonian Era is characterized by a thin atmosphere, dry and dusty conditions, freezing surface temperatures, and occasional water releases, probably from localized ice melting. Volcanism and debris impacts continue but at a significantly lower rate than in the previous eras.

This 4.5 billion-year-old rock, labeled meteorite ALH84001, is believed to have once been a part of Mars and to contain fossil evidence that primitive life may have existed on Mars more than 3.6 billion years ago. The rock is a portion of a meteorite that was dislodged from Mars by a huge impact about 16 million years ago and that fell to Earth in Antarctica 13,000 years ago. The meteorite was found in Allan Hills ice field, Antarctica, by an annual expedition of the National Science Foundation's Antarctic Meteorite Program in 1984.

ALH84001,0

This simulated view shows Mars as it might have appeared during the height of a possible ice age in recent times, geologically speaking.

Of all Solar System planets, Mars has the climate most like that of Earth. Both are sensitive to small changes in orbit and tilt. During a period from about 2.1 million to 400,000 years ago, an increased tilt of Mars' rotational axis caused increased solar heating at the poles. This polar warming may have caused water vapor and dust to be thrown into the atmosphere and deposits of ice and dust to form on the surface down to about 30 degrees latitude in both hemispheres. That is the equivalent of the southern United States or Saudi Arabia on Earth. Mars has been in an interglacial period characterized by less axial tilt for about the last 300,000 years.

In this illustration, prepared for the December 18, 2003, cover of the journal Nature, the simulated surface deposit is super-imposed on a topography map based on altitude measurements taken by Mars Global Surveyor and images from NASA's Viking orbiters of the 1970s.

Because of the length of the orbit, Martian seasons last about twice as long as Earth's. However, the seasons themselves on Mars are unequal. Due to Mars's orbital eccentricity, summer in the southern hemisphere (when the south pole tilts towards the Sun) occurs during perihelion, when Mars is moving more rapidly in its orbit. This results in a shorter southern hemisphere summer than the summer in the northern hemisphere, which occurs during aphelion, when Mars is moving slower. The lengths of winter reverse with a northern winter that is shorter than winter in the south.

Remarkably, the axial tilt of Mars is changeable. Mars does not have a massive satellite, like Earth does, to stabilize its rotation and fix the axial tilt. It does have two moons, but they are irregular-shaped asteroid-sized satellites about 27 and 15 km (17 and 9.3 m) at their widest points. The two moons, Phobos and Deimos, the Roman words for fear and dread, whip around the planet in orbits lasting only 7.6 and 30.2 hours, respectively. Without a stabilizing moon, Mars's axial tilt can shift to as little as 0 or to as high as 40°. Higher axial tilts would cause Mars's poles to be bathed in 300 or more days of continuous sunlight during each orbit. Polar heating would result in a nearly complete release of polar water and carbon dioxide to the atmosphere. The impact on the climate and the surface features of Mars would be considerable.

Navigating Mars

Sometimes, what seems like a good idea comes back to haunt us. The United States is still locked in the cumbersome British System of measurements, and geographic longitude is confused by east and west directions and a dateline of 180°east or west of the Prime Meridian. With no politicians lobbying for the Prime Meridian to run through their countries, the coordinate system on Mars was set up the right way from the beginning. Like Earth, Mars has an equator based on its axis of rotation (halfway between the north and south poles). It has a prime meridian that intersects the equator and runs pole to pole. The location of the prime meridian was chosen to run through the oldest terrain of Mars—Noachian Terra. Longitude numbers begin at the prime meridian (0°) and increase westward until the planet is circumnavigated back to the starting point (360°). The location of any feature on Mars is described as 0° to 90° north or south latitude and 0° to 360° longitude.

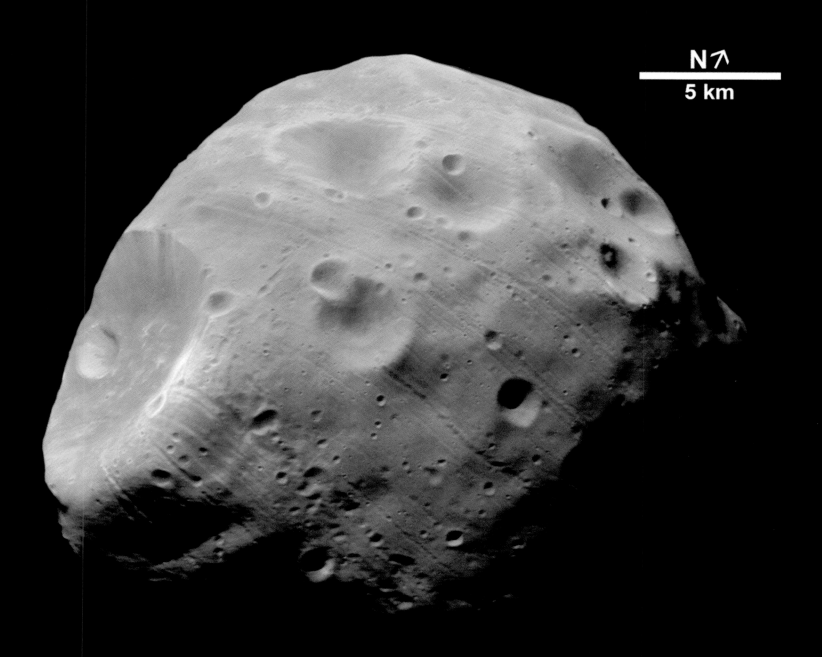

In spite of climatic variations, Mars has been a relatively stable planet over eons of time. On Earth major crustal movements and wind, water, ice, chemical, and biological erosion continually remodel Earth's surface. Land is built up and land is torn down. Consequently, most of Earth's exposed land surface is less than 1 billion years old, and outcrops more than 3 billion years old are exceedingly rare.

On Mars the pace of change is slow, but like the tortoise and the hare fable, the cumulative results are dramatic. Mars has a full range of surface ages stretching from the very young to the very old—4 billion or more years. It preserves many of its impacts, especially in the southern highlands. On Earth, the surface effects of impacts have largely been obliterated by erosion.

This image, taken by the High Resolution Stereo Camera (HRSC) on board ESA's Mars Express spacecraft, is one of the highest-resolution pictures so far of the Martian moon Phobos. It shows the Mars-facing side of the moon, taken from a distance of less than 200 kilometers.

This artist's impression shows ESA's
Mars Express spacecraft scanning the
fast-moving shadow of the moon
Phobos as it traveled across the
Martian surface.

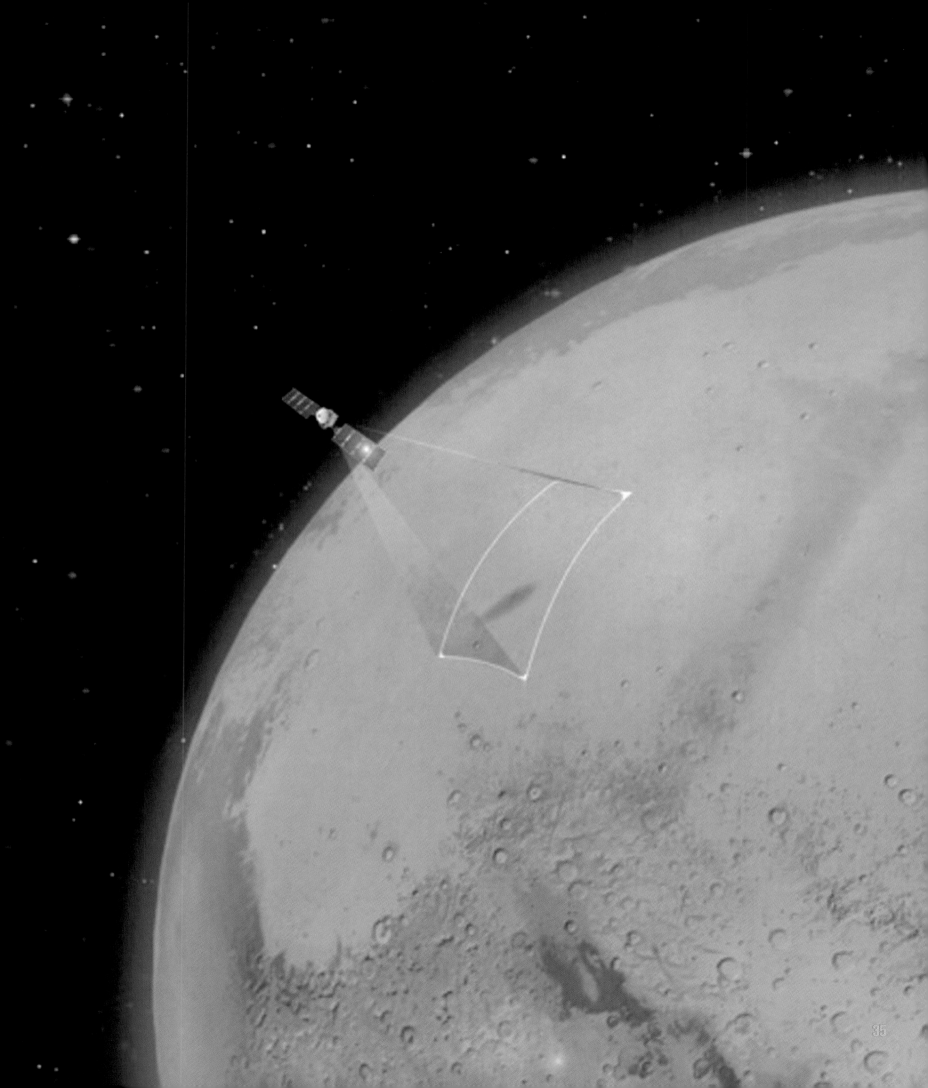

Images from Mariner 9 in 1972 revealed that some of the mesas and mounds found within the chasms of the Martian "Grand Canyon"—the Valles Marineris—have layers in them. Speculations as to the origin of these interior layered materials ranged from volcanic ash deposits to sediments laid down in lakes that could have partially filled the Vallis Marineris troughs, much as lakes now occupy portions of the rift valleys of eastern Africa. The proposal that the Valles Marineris once hosted deep Martian lakes led to speculation about the possibilities of finding fossil evidence of Martian life.

Mars Global Surveyor (MGS) Mars Orbiter Camera (MOC) images have ten or more times better resolution than the Mariner 9 and Viking orbiter images taken in the 1970s. MOC images have indeed confirmed the presence of layered outcrops within the Valles Marineris. The layered rock is now visible because of faulting and erosion. The high-resolution picture shown here was the first image received by MOC scientists that began to hint at a larger story of layered sedimentary rock on Mars. The MOC picture shows an area in far southwestern Candor Chasma that was not previously expected to exhibit layers.

Layers indicate change. The uniform pattern seen here—beds of similar properties and thickness repeated over a hundred times—suggest that the deposition of materials that made the layers was interrupted at regular intervals. Patterns like this, when found on Earth, usually indicate the presence of certain underwater environments. On

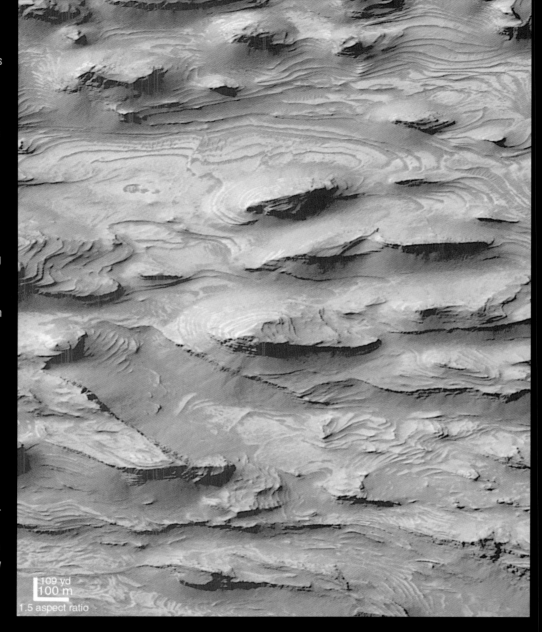

109 yd
100 m
1.5 aspect ratio

Mars, these same patterns could indicate that the materials were deposited in a lake or shallow sea. Other MOC images suggest that these layers would not have formed in a lake, but instead were deposited in a crater or other basin that existed before the chasm was cut into the surrounding terrain, watery or otherwise.

Tectonic fractures within the Candor Chasma region of Valles Marineris retain ridge-like shapes as the surrounding bedrock erodes away. This points to past episodes of fluid alteration along the fractures and reveals clues into past fluid flow and geochemical conditions below the surface.

The High Resolution Imaging Science Experiment camera on NASA's Mars Reconnaissance Orbiter took this image on Dec. 2, 2006. The image is approximately 1 kilometer (0.6 mile) across.

The crust of Mars is not broken into movable tectonic plates, as is Earth's crust. Movements are primarily up and down rather than horizontal. Its long-term stability and lower gravitational attraction has resulted in the buildup of huge volcanic edifices dwarfing similar features on Earth. The mass of these features has stressed the nearby crust with many small and large rifts that have widened through landslides and surface collapse.

We All Have Our Problems

The news is full of stories about global warming and what planet-wide changes might mean to future generations on Earth. One of the advantages of studying other planets is that scientists can observe the effects of widely differing conditions without using Earth as a test tube. A very dense carbon dioxide-rich atmosphere on Venus and its proximity to the Sun have led to runaway global warming. Mars, with very little atmosphere and a far greater distance from the Sun, is a frigid world. No global warming there. Well, not quite. Global warming doesn't mean a planet gets hot. It just gets warmer than normal. Slight temperature changes can lead to major surface changes. Dust on Mars appears to be triggering a global warming trend that has warmed the planet by 1.17° C (2.1° F) over the last 20 years. Strong winds on Mars can sweep light-colored dust off the surface and make local changes in the albedo, or reflectivity. Darker surface rocks absorb more heat from the Sun, and some of this heat is transferred to the atmosphere. Like on Earth, unequal heating leads to atmospheric circulation—wind and storms. This leads to blowing dust, more heating, stronger winds, more dust, etc. Computer weather models indicate the cycle will continue, and one of its effects will be increased Martian polar cap melting. Sounds like Earth!

Erosion on Mars is apparent. Sedimentary rock layers that extend for great distances are especially visible in the northern lowlands. Martian winds have sculpted rock and produced light and dark surface streaks and wind shadows that indicate wind direction. Sand dunes cross open plains and fill in crater floors. Streaks, caused by dust devils, can be seen from orbit. Although liquid water on the surface is rare, there is strong indication of past wetter times in the many braided channels preserved along slopes. Currently, the dominant surface-shaping forces on Mars are wind, landslides, meteorite and asteroid impacts, and volcanism.

These images, taken by the High Resolution Stereo Camera (HRSC) on board ESA's Mars Express, are of the Acheron Fossae region, an area of intensive tectonic (continental 'plate') activity in the past. The images show traces of enormous stress and corresponding strain in the crust of the Red Planet. The feature is situated about 1,000 kilometers north of the large Olympus Mons volcano. .

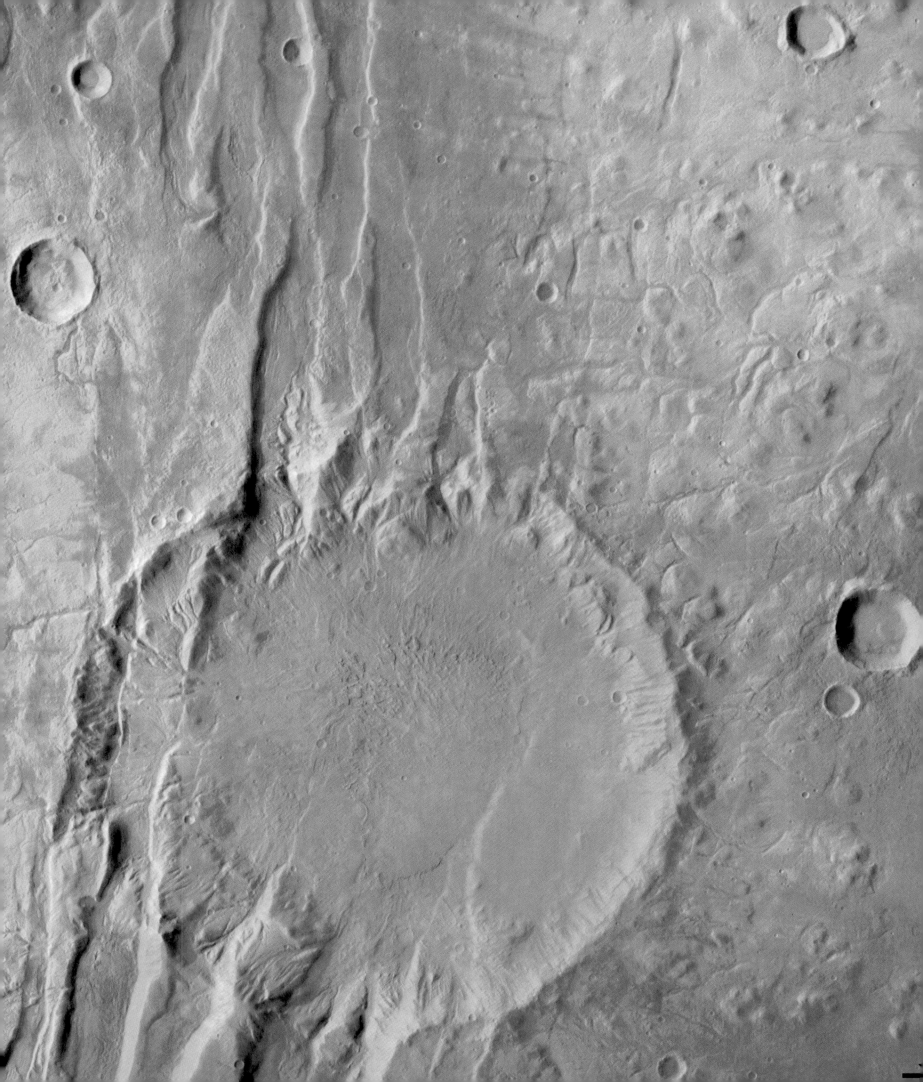

This picture, from March 2005, is a composite of daily images acquired by the Mars Global Surveyor during the previous year. The picture shows the Syrtis Major face of Mars when it is spring in the south and autumn in the north.

2 TO HELLAS AND BACK

The orbit of Mars keeps the planet just out of clear viewing range. Even with the best Earth telescopes, Mars is still a blurry world of light and dark. Much is left to the imagination. Regardless, observers of a century ago constructed fantastic surface maps dotted with amazing features that have mostly disappeared with clearer views from orbital spacecraft. However, some of the largest features survived. In 1659, Dutch astronomer Christian Huygens sketched a huge triangular-shaped shadowy feature in the mid-Martian latitudes. Over time, he improved on his sketches and added a bright circular area at the South Pole. The circular feature was likely the first observation of the Martian south polar cap.

Using an early telescope, astronomer Christian Huygens created one of the first sketch maps of the Martian surface. In 1659 he drew shadowy surface features that are thought to be the first record of the Syrtis Major region of Mars.

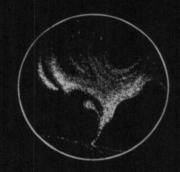

Willian Rutter Dawes was another of the nineteenth-century Mars observers that mapped the Martian surface. Here is one of his representations of Syrtis Major.

Over the centuries, maps of the triangular patch, christened Syrtis Major, became less sketchy and more refined, with a graceful curve that swept towards the west. Observers noted that the darkness of the patch and its size seemed to vary with the seasons. They speculated that the patch grew and darkened because Syrtis Major was covered with vegetation. Vegetation was thin and light in the Martian spring but dark and lush in the fall.

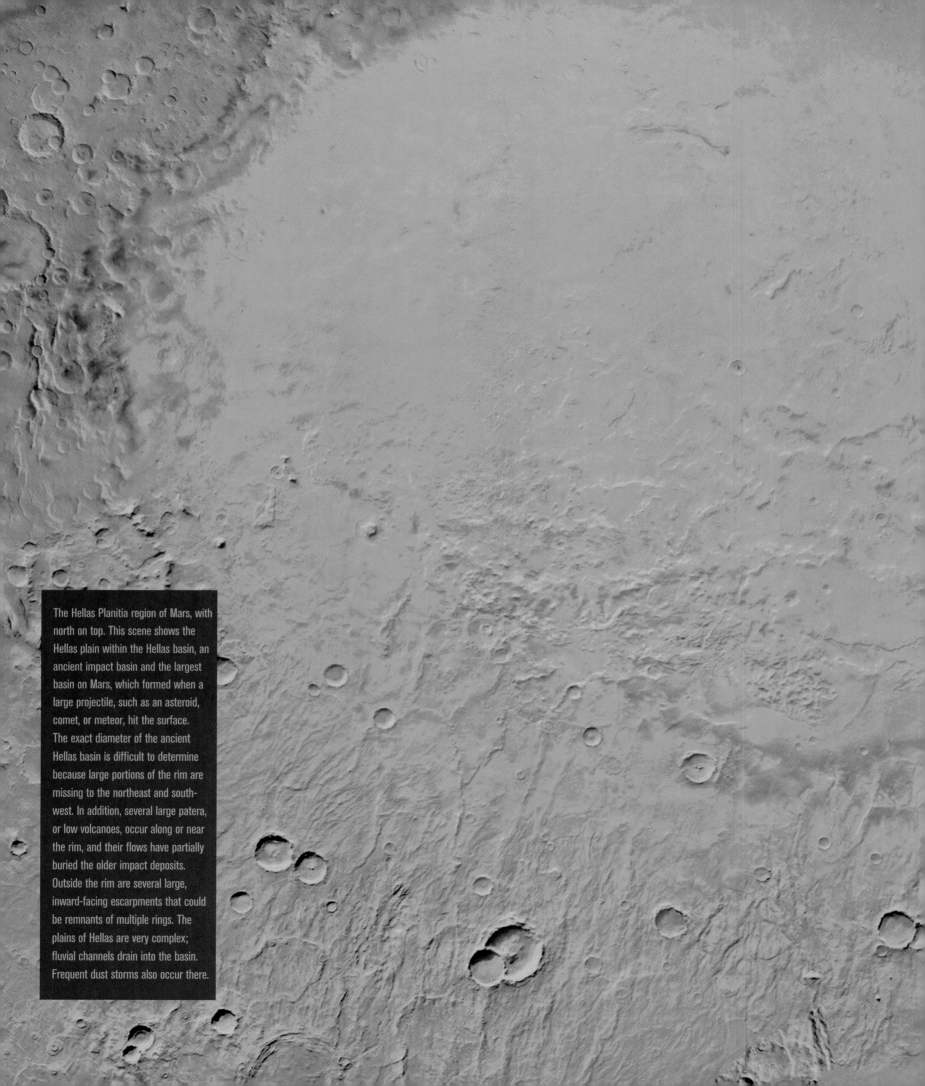

The Hellas Planitia region of Mars, with north on top. This scene shows the Hellas plain within the Hellas basin, an ancient impact basin and the largest basin on Mars, which formed when a large projectile, such as an asteroid, comet, or meteor, hit the surface. The exact diameter of the ancient Hellas basin is difficult to determine because large portions of the rim are missing to the northeast and southwest. In addition, several large patera, or low volcanoes, occur along or near the rim, and their flows have partially buried the older impact deposits. Outside the rim are several large, inward-facing escarpments that could be remnants of multiple rings. The plains of Hellas are very complex; fluvial channels drain into the basin. Frequent dust storms also occur there.

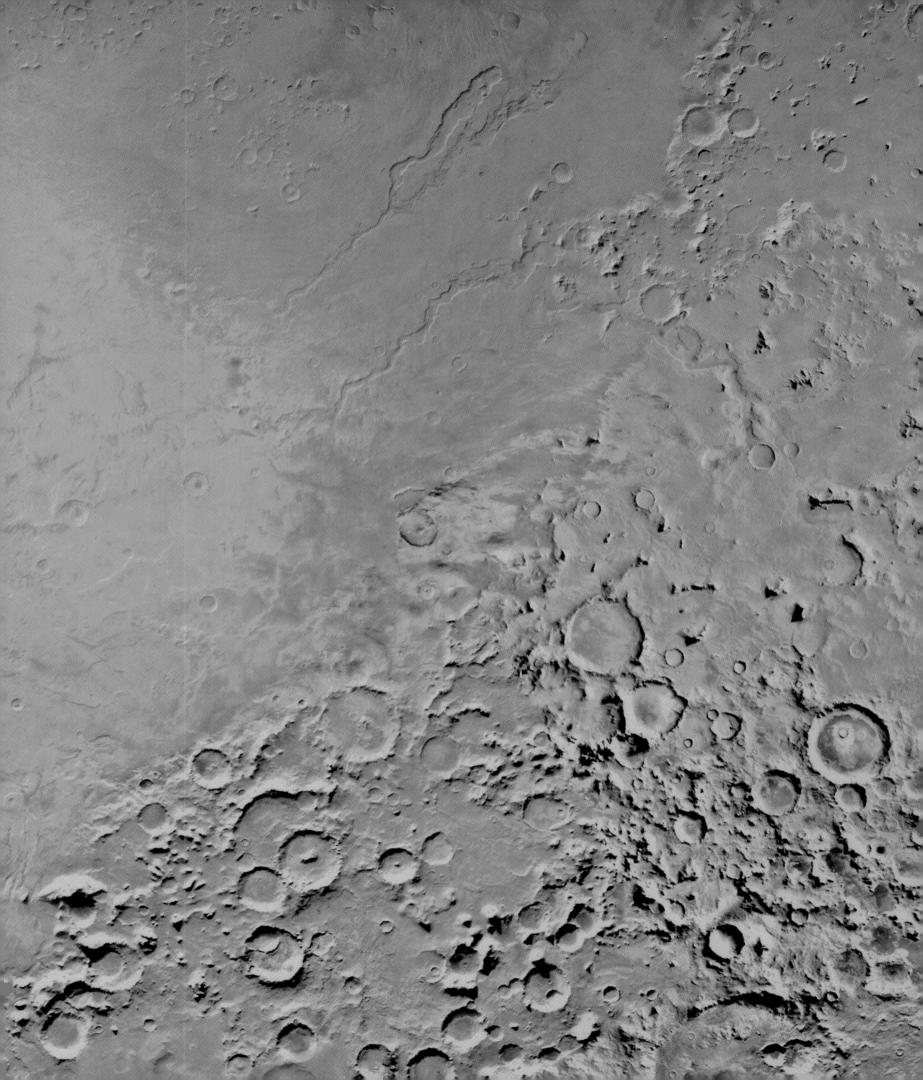

Abruptly, the vegetation hypothesis gave way to data collected from orbital spacecraft. Like much of the surface of Mars, Syrtis Major is covered with basaltic lava similar to the basalts of Earth. Laser altimeter measurements enabled scientists to determine that Syrtis Major is actually a low conical volcano that spans 1,200 km (746 m) and rises 2 km (1.2 m) to its central caldera. The light and dark of Syrtis Major are due to weathering effects. Basalt, a dark gray volcanic rock, weathers to a lighter rusty brown. The weathered surface flakes off and alters, especially in the presence of moisture, to a still lighter-shaded dust. Seasonally, Martian winds transport the dust off the gentle slopes of Syrtis Major, exposing fresh basalt and making its surface appear darker again. Further weathering rebuilds the dust coating in a cycle accounting for the brightness changes once attributed to vegetation.

There were other Martian features visible from Earth, but again not with enough clarity to be certain of their nature. One was a large circular patch directly south of Syrtis Major. In early charts, the patch was sketched as an area brighter than the surrounding land. Observers noted that the patch was typically brighter in the morning and dimmed as the Martian day wore on.

It wasn't until the Mariner 9 spacecraft orbited Mars that the bright circular patch was revealed to be a huge impact basin with a diameter of about 1,700 km (1,056 m). In its early days, Mars was struck with an asteroid that generated enough kinetic energy to blast out a crater large enough to cover the state of Alaska. The floor of the crater, properly identified as an impact basin because of its size, is a relatively flat plain that lies 8.2 km (5.1 m) below the surrounding uplands. It is the lowest region of Mars. The feature was named Hellas Planitia by Giovani Schiaparelli, after the ancient Greek name for Greece.

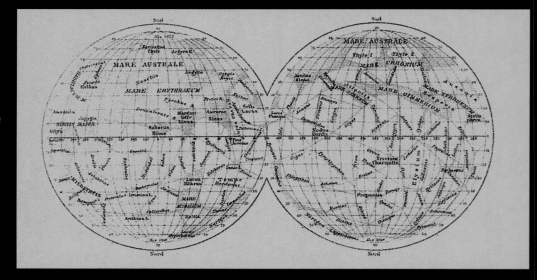

With Mars in a favorable position in the sky, Giovanni Schiaparelli created a highly detailed map of the Martian surface. Numerous straight lines, he called canali, crisscross the surface. This map, based on the 1877 observations, was the starting point for a 100-year-long debate on whether or not Mars was inhabited by intelligent life.

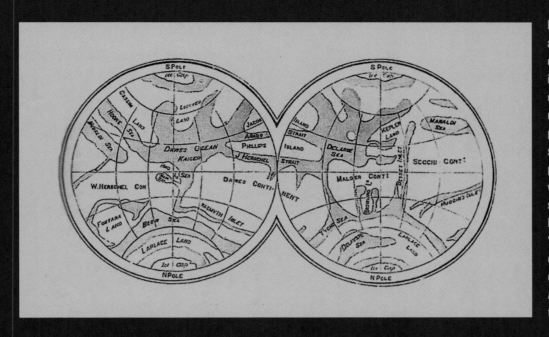

Richard Proctor made the first systematic attempt to create a nomenclature of the surface features of Mars. He identified oceans and continents and curious ring-shaped water bodies surrounding the Martian poles. This map, published in 1871, shows the Martian south pole at the top and the north pole at the bottom. This is an accurate view of what he saw through his telescope. Astronomical telescopes typically invert and reverse images. "Correcting" images require additional optics but diminish the optical quality of the image in the process. Astronomers prefer better image quality to corrected displays.

This is a regional topographic model of the Hellas basin. The Hellas basin, the lowest place on Mars, shows its profile in the diagram above the false-colored image. The elevations of the profile are highly exaggerated to show detail. Reds and yellows in the image mark the highest elevations of the basin and the blue and violet its depths. The black line indicates where the average or zero elevation of Mars intersects the basin.

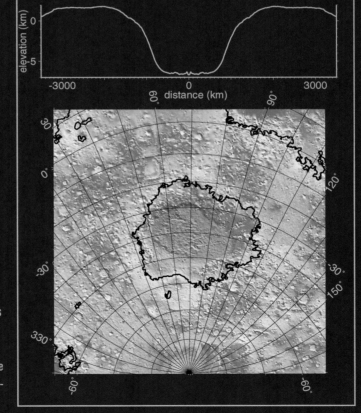

Dating the Hellas impact is somewhat difficult, but clues abound. The circular edge of Hellas should be a mountainous rim dropping steeply towards the impact center and sloping more gradually to the outside. Instead, the rim is characterized by low hills intersected by rough valleys that indicate a long period of wear. The debris, mounded up around the basin by the explosion, consisted of rubble that was easily eroded. Ancient drainage channels intersect the rim and basin floor, indicating wetter times on Mars. Presumably, the rim was worn down long ago by the action of water and wind.

Another age-dating clue came from magnetometer studies of Mars. Mars is almost devoid of a dipolar magnetic field. In other words, future Martian explorers would find a compass useless. Earth has a planet-wide magnetic field that is generated by the slow movement of molten iron in its outer core. A lack of an easily detected dipolar field indicates that the core of Mars is largely solid. However, it wasn't that way in Mars's early days.

The young Mars had an active core and an accompanying magnetic field. However, unlike Earth, it did not have the size necessary to sustain internal heat and core movements over 4 1/2 billion years. Today, only remnants of its magnetic field remain. The remnants are the north-south alignments of magnetic minerals (like tiny compass needles frozen in place) that occurred as early Martian rocks solidified and are detectable in the southern highlands but are wholly absent on the floor of the Hellas basin. It appears the impactor that created the basin blasted a hole in the magnetic highlands after Mars's dipolar field had faded. Consequently, magnetic mineral alignments in the basin were not reestablished in the cooling that followed the impact.

Details in a fan-shaped deposit discovered by NASA's Mars Global Surveyor orbiter provide evidence that some ancient rivers on Mars flowed for a long time, not just in brief, intense floods.

Today, the fan is today a deposit of sedimentary rock. The fan's general shape, the pattern of its channels, and its low slopes provide circumstantial evidence that the feature was an actual delta, that is, a deposit made when a river or stream enters a body of water. If so, this landform is a strong indicator that some craters and basins on Mars once held lakes.

The picture is a mosaic of images acquired between August 2000 and September 2003. North is up. Sunlight illuminates the scene from the left. The fan is in an unnamed crater that is 64 km (40 miles) in diameter. The crater lies northeast of a larger one named Holden Crater.

The fan is a fossil landform. That is, it is an eroded remnant of a somewhat larger and thicker deposit. The originally loose sediment was turned to rock and then eroded over time. The channels through which sediment was transported are no longer present. Instead, only their floors remain, and these have been elevated by erosion so that former channels now stand as ridges.

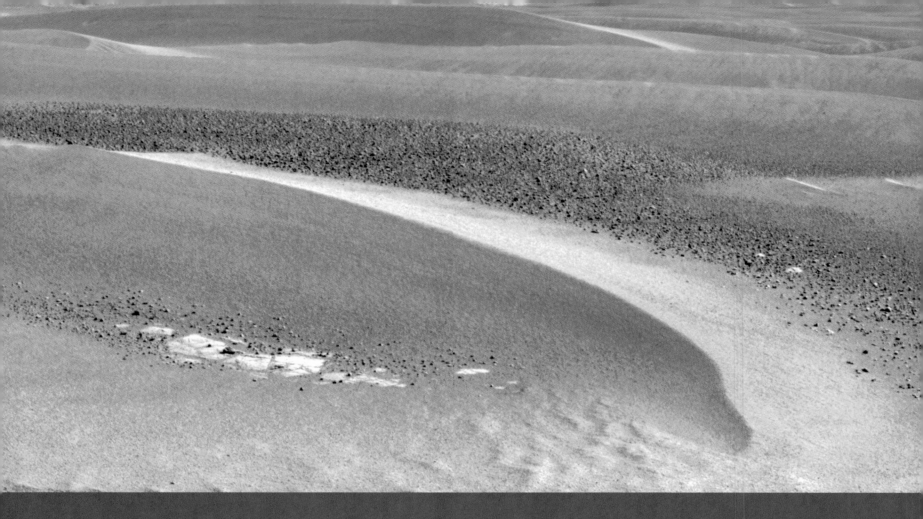

As the rover Opportunity continues to traverse from Erebus Crater toward Victoria Crater, it navigates along exposures of bedrock between large, wind-blown ripples. Along the way, scientists have been studying fields of cobbles that sometimes appear on trough floors between ripples. They have also been studying the banding patterns seen in large ripples.

The origins of cobble fields like this one are unknown. The cobbles may be a lag of coarser material left behind from one or more soil deposits whose finer particles have blown away. Or, they may be eroded fragments of meteoritic material, secondary ejecta of Mars rock thrown here from craters elsewhere on the surface, weathering remnants of locally derived bedrock, or a mixture of these. This is a false-color rendering that is used to enhance differences between types of materials in the rocks and soil.

The most obvious indicator of ancient age is the location of the basin. The basin overlies the ancient southern cratered highlands. The impact is therefore likely to be younger than, or at least coming at the end of, the great episode of intensive meteorite and asteroid bombardment that characterized the early days of Solar System formation. Strong evidence of the bombardment is clearly visible on the Moon and Mercury. (Evidence of the bombardment on Earth is minimal, due to the continuous active remodeling of the surface with volcanism, plate tectonics, and weathering.) On Mars, this early period is known as the Noachim era (4.5 to .5 billion years ago). The Hellas impact probably took place between 3.8 and 3.7 billion years ago.

With its severely eroded rim, one would expect the depressed floor of the Hellas basin to be filled with wind-blown sand and dust, resembling a giant sand trap on a golf course. It is partially filled with sediment. However, the covering is uneven, and sedimentary layering is distinctly visible. Wind or water-transported sediment was deposited in the basin over long periods of time and eroded during other intervals.

The far southern latitude of the Hellas Planitia and its low topographic altitude makes it one of the coldest locations on Mars. During the mornings, especially, the floor may exhibit a brilliant white coating of carbon dioxide snow that sublimates back into the atmosphere during the warming of the day. This accounts for the brightness of the basin observed long ago. The basin of Hellas Planitia is a fairyland of swirls, dunes, layers, bedrock, and temporal frost

Rates of change in surface temperatures during a Martian day indicate differences in particle size in and near Bonneville Crater. Temperature information from the miniature thermal emission spectrometer on NASA's Mars exploration rover Spirit is overlaid onto a view of the site from Spirit's panoramic camera.

In this color-coded map, red during the day suggests sand or dust. (Red is about 270 Kelvin, or 27° Fahrenheit.) An example of this is in the shallow depression in the right foreground. Areas that stay blue longer into the day have larger rocks. (Blue indicates about 230 Kelvin, or -45° F.) An example is the rock in the left foreground.

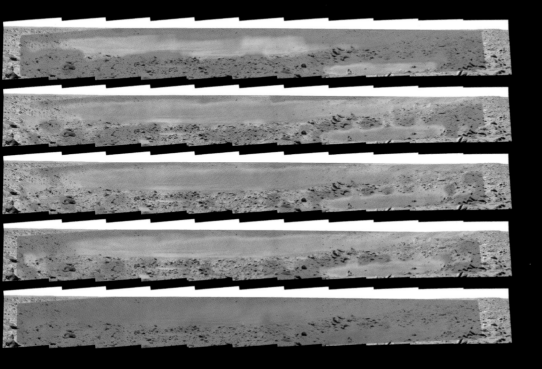

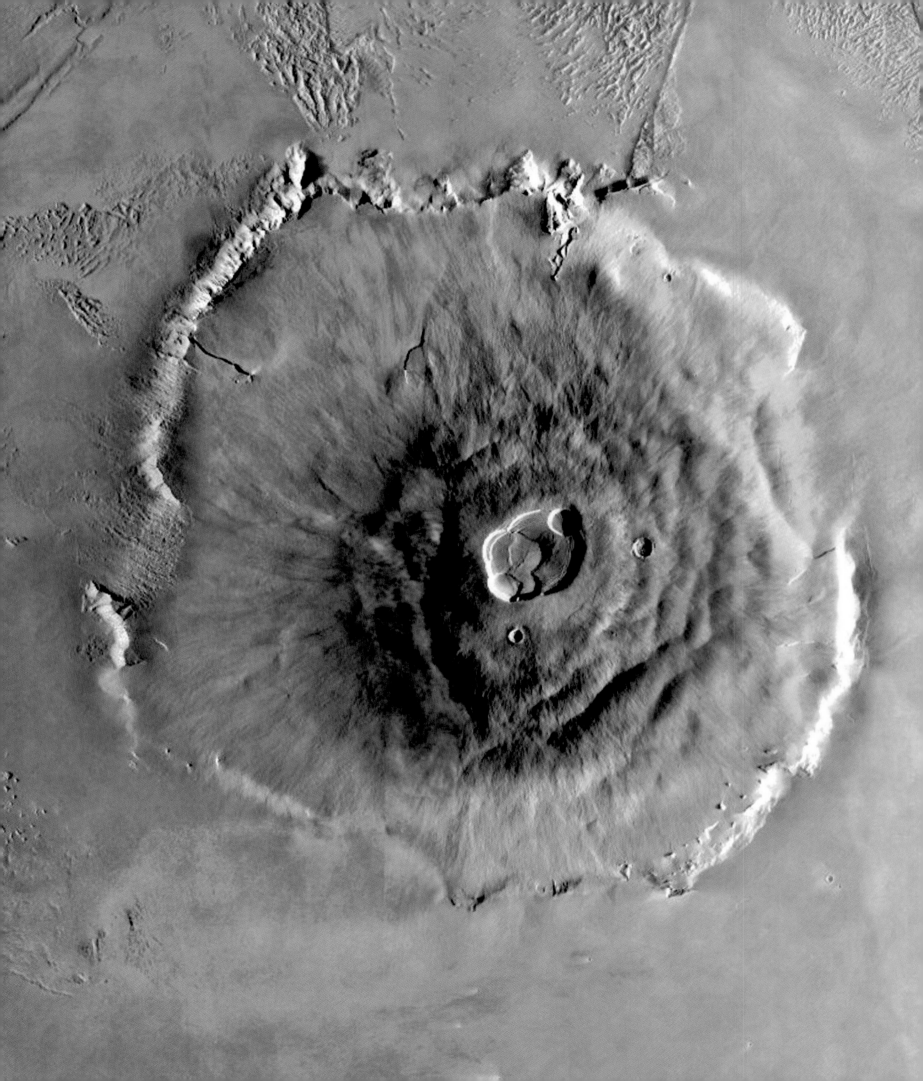

SNOWS OF OLYMPUS

Thankfully, the Mariner 9 Mars probe was an orbital mission. Had it been another flyby spacecraft, its cameras would only have recorded pictures of dust. The planet-wide dust storm Mariner 9 encountered, as it slipped into orbit, obscured the entire surface. Except for a few dark patches, Mars appeared featureless. But, intriguingly, those dark patches matched the coordinates of an enduring Martian mystery.

In the 1800s, astronomers noticed persistent but variable bright spots on the Martian surface. The spots were located in the broad Tharsis region. Three of the spots formed a straight line that intersected and straddled both sides of the Martian equator at approximately 110 degrees longitude. Another spot was 20 degrees further due north. The largest and brightest spot was slightly to the west, at 130 longitude, 20 degrees north latitude. Giovani Schiaparelli, soon to be famous for his Martian canal observations, named it Nix Olympica, after the classical Greek snow-capped home of the gods.

For more than a hundred years, Martian astronomers had observed daily brightness changes in Nix Olympica. On occasion, the spot appeared to be ring-shaped. Normally, brightening occurred in the early afternoon, Martian time. By evening, the spot would rival the brightness of the polar caps. Next morning, the spot would be dim, but it would grow in brightness with the advance of the day.

Color mosaic of Olympus Mons volcano on Mars from Viking 1. Olympus Mons is about 600 km in diameter and its summit is 24 km above the surrounding plains. (Viking 1 Orbiter MH20N133-735A)

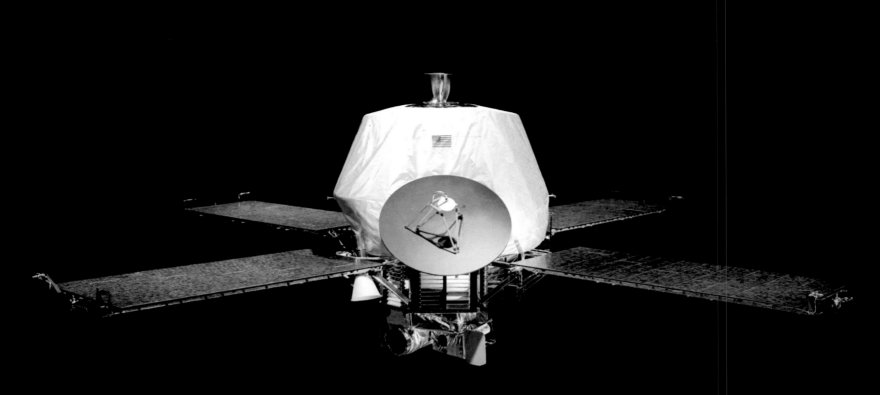

The Mariner Mars mission was planned for two identical spacecraft. Unfortunately, Mariner 8 was destroyed because of a launch failure. Mariner 9 had to go it alone. It was launched on May 30, 1971, and traveled 398 million km before reaching Martian orbit. Instrumentation was housed in an octagonal box, and power was provided by four solar panels stretching almost 7 m tip to tip. During its mission, the spacecraft gathered more than 7,300 pictures and made stunning discoveries of the Red Planet.

By late November 1971, the planet-wide dust storm started to settle. Nix Olympica began coming into focus, revealing a complex structure of volcanic collapse calderas and making clear to scientists that this Martian feature was far more interesting than the earlier ring-shape structure revealed in telescopes and pictures from earlier spacecraft. Nix Olympica was later renamed Olympus Mons.

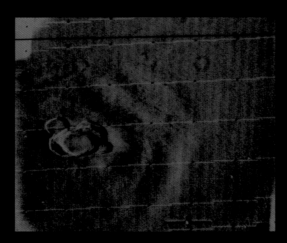

In 1907, E.C. Slipher of the Arizona Lowell Observatory suggested that the brightening of Nix Olympica and the other nearby spots was due to the formation of clouds. It was a logical hypothesis that could not be verified with the technology of the day. The cause of the brightness change was still an open question when Mariner 9 arrived.

Toward the close of January 1972, the raging planetary dust storm had all but ended. In a single image sent back by Mariner 9, the bright spot mystery was solved. The photograph revealed large craters centered on each spot. By comparing the craters to the structures of terrestrial and lunar craters, the answer to the mystery became obvious. The craters were collapsed calderas created at the end of a volcanic eruption, when supplies of magma became exhausted. Nix Olympica and the other nearby bright spots were huge volcanoes so high their summits had darkened the upper reaches of the dust storm.

This mosaic of global color images features the Tharsis volcanoes (mostly covered by bluish-white water ice clouds) and the Valles Marineris trough system (to the right). The images were obtained on a single Martian day.

The mysterious brightness changes were indeed due to cloud formations. Clouds formed daily as the thin Martian atmosphere flowed over the volcanoes, cooled, and condensed. The occasional observed bright ring shapes were due to ground fog that sometimes collected around the volcano's flanks. With its true nature understood, Nix Olympica was renamed Olympus Mons (Mount Olympus). The other volcanoes are Arsia Mons, Pavonis Mons, and Ascraeus Mons, following a straight south to north fault line, and the distant northern volcano was named Alba Patera. A patera is a shallow, scalloped crater, usually a volcanic caldera. Alba Patera, at 110 degree longitude, 40 degree north latitude, is a much shallower volcanic edifice than the other four volcanoes described, but it has a much broader base and a huge system of concentric cracks large enough to cover the state of Texas.

This close-up view, looking southwest, shows a perspective of the eastern scarp of the Olympus Mons volcano.

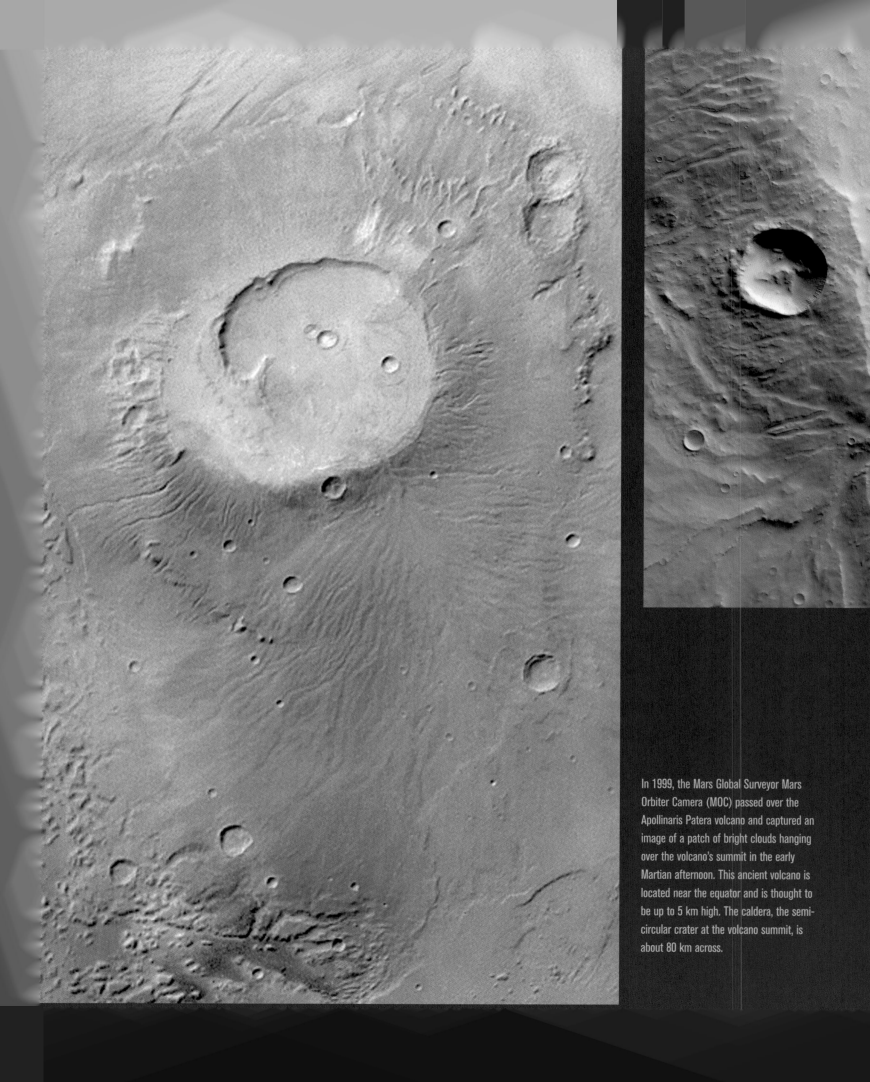

In 1999, the Mars Global Surveyor Mars Orbiter Camera (MOC) passed over the Apollinaris Patera volcano and captured an image of a patch of bright clouds hanging over the volcano's summit in the early Martian afternoon. This ancient volcano is located near the equator and is thought to be up to 5 km high. The caldera, the semi-circular crater at the volcano summit, is about 80 km across.

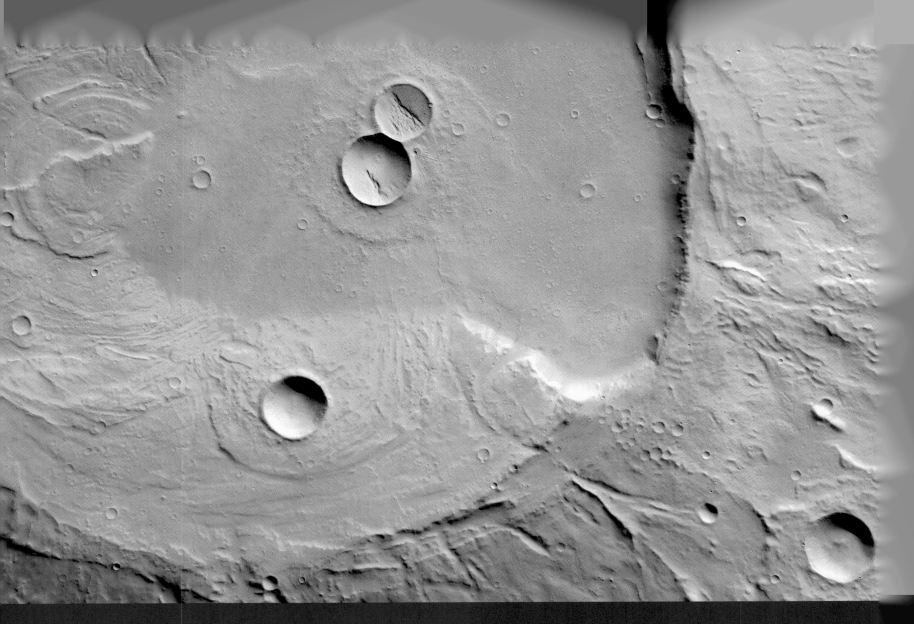

Seen here is the caldera of Apollinaris Patera, an ancient 5-km-high volcano northwest of Gusev Crater. In this true-color image, the terrain is partly covered by thin, diffuse, whitish-appearing clouds. North is to the right.

Because of orbital mapping, we have learned that the Tharsis region, home to Olympus Mons, four other massive volcanoes, and a few hundred smaller volcanic edifices, is a broad volcanic uplift dome about 3,000 km (1,900 m) across. The main part of Tharsis is large enough to cover all of Europe. From space, Tharsis appears as a broad plain, but it is unlike any plain on Earth. Its central bulge pushes up about 9 km (5.6 m) above the average elevation of Mars.—higher than Mount Everest on Earth.

Geologists suggest that the initial buildup of the Tharsis dome, containing approximately 300 million cubic km of lava, took place during the Noachian era perhaps 4 billion years ago. The volcanoes on top of the dome are more recent and are thought to be dormant. That means the Tharsis region represents potentially 4 billion years of continuous volcanic activity.

Although Earth, Venus, and Jupiter's moon Io all experience volcanic activity, there is nothing on these worlds that comes close to Mars's volcanoes and Tharsis. Earth's biggest volcano, Mauna Loa on the island of Hawaii, is 8 km (5 m) high if you measure up from the seafloor. Olympus Mons rises 21 km (13 m) from the Tharsis plain, and its bulk mass is about three times greater than that of Mauna Loa. Even Arsia, Pavonis, and Ascraeus Mons, at 17.8, 14.1, and 18.2 km (11.1, 8.8, and 11.3 m), respectively, dwarf Mauna Loa in height and bulk mass.

The base of Olympus Mons is circular and 600 kms across. It is clearly marked with a 1-km (0.6-m)-high escarpment. There is no known parallel on Earth to the Olympus Mons cliffs; other volcanoes do not have them. How and why the cliffs formed may be a question Mars-bound astronauts will have to answer.

One would think a glimpse of Olympus Mons from the Martian surface would be one of the great sights in the Solar System. Surprisingly, this isn't so. The volcano's broad base and Mars's rapid curvature, due to the planet's small 6,780-km (4,213-m) diameter, would prevent a future astronaut-tourist from seeing the summit while standing near its base. The best view of the volcano would be from an orbiting spacecraft, where the whole of the mountain could be viewed.

Imagine the journey of a future Mars explorer moving up the gradual slope of Olympus Mons. Along the way, probably riding in some sort of rover, the explorer would mount many overlapping lava flows, each like the page in a book on the history of the mountain, and cross many hardened lava channels and collapsed lava tubes In a traverse of more than 280 km (175 m) and an elevation gain of 20 km (12 m), the summit would be reached.

At the summit the explorer would behold a 40-km (24.8-m)-wide caldera, which is large enough to swallow Los Angeles. This caldera is a complex structure, with many smaller intersecting calderas that formed following recurrent eruptions, when empty magma chambers collapsed in on themselves. The main caldera floor, 2 km (1.2 m) below the rim, is a jumble of wind-blown dust and sand dunes, scoured rocky plains, lava channels, cracks, rifts, and lava tubes.

One thing the explorer would notice during the volcano traverse is the absence of significant impact craters. Satellite views show the caldera floor and the mountain's flanks remarkably smooth. The surface of Olympus Mons looks fresh. Any large impacts that may have occurred on the mountain have been covered up by lava flows as recent (at least by geological standards) as 10 million years ago. Olympus Mons, a dormant volcano, most likely will erupt again in the future.

The massive Tharsis province and its volcanoes say much about the nature of Mars. A volcanic dome with Mars-sized volcanoes resting on top is not possible on Earth for a couple of reasons. Due to Earth's greater mass and density, Earth's surface gravity is 2.6 times the surface gravity on Mars. Earth's crust simply could not

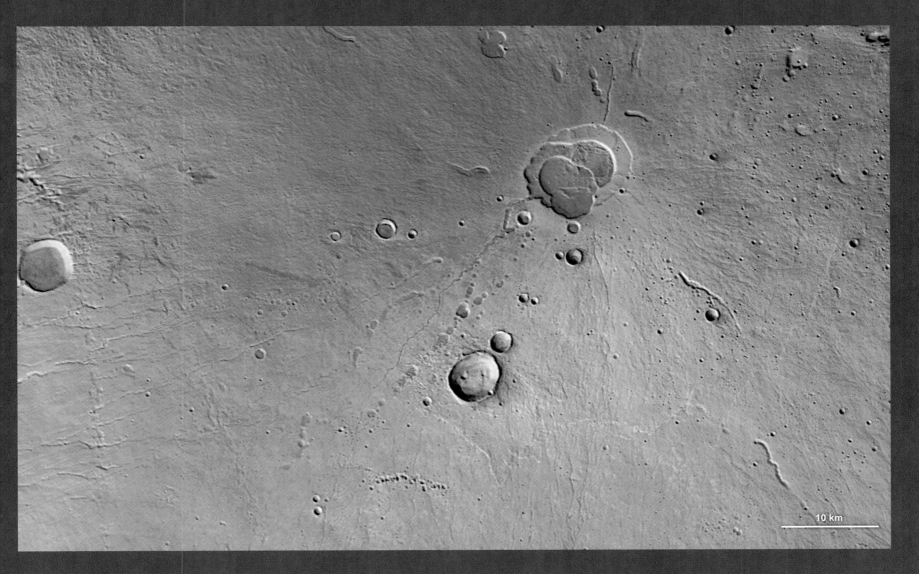

This image of the Hecates Tholus volcano was taken by the High Resolution Stereo Camera (HRSC) on Mars Express in orbit and from an altitude of 275 km. It shows the summit caldera of Hecates Tholus, the northernmost volcano of the Elysium volcano group. The volcano reveals multiple caldera collapses. On the flanks of Hecates Tholus several flow features related to water (lines radiating outwards) and pit chains related to lava flow can be observed. The volcano has an elevation of 5,300 m; the caldera has a maximum diameter of 10 km and a depth of 600 m.

support a Tharsis mass and its volcanoes. The volcanic rock and the underlying crust would be driven into the mantle for remelting.

Another important factor limiting the height of Earth's volcanoes is the mobility of Earth's crust. Earth's crust is broken into about 15 or so continent-sized and smaller tectonic plates that slowly shift position relative to each other. Evidence of the shifting is clearly seen in the Hawaiian Island chain in the Pacific Ocean. The islands string out in a west northwest direction as the Pacific plate slowly moves across a persistent hot spot in Earth's mantle. The huge Tharsis bulge and its volcanoes indicate that Mars has a thicker and more stable crust that has remained in place for billions of years.

The Valles Marineris canyon system is over
3,000 km long and up to 8 km deep,
extending from Noctis Labyrinthus, the
graben system to the west, to the chaotic
terrain to the east.

4 THE GRANDEST CANYON

With the abatement of the 1971 planet-wide dust storm, a small, previously known dark-ish region, to the east of the Tharsis volcanic dome, changed character. This region had been identified as an insignificant "canal" on a Mars map drawn by Percival Lowell. He named the canal Coprates, after an ancient Persian river.

Rather than darkening, as the giant Tharis volcanoes did when they emerged from the storm, Coprates morphed into a bright, intricate, sinuous pattern that stretched eastward across the Tharsis dome and hooked northward into the Chryse Planitia. It soon became apparent that Coprates was a canyon structure draped in a light-colored residual dust that continued to hover above the canyon floor.

As the dust in the canyon settled, Mariner 9 scientists were stunned. This was no ordinary canyon. It was (and still is) the largest known canyon system anywhere in the Solar System. On Earth, this canyon would stretch from New York to Los Angeles—about four times longer than Arizona's Grand Canyon. In some places, it is over four times as deep. The Grand Canyon is more comparable in size and depth to one of the side canyons, a tributary, of the Martian canyon. Mariner 9 scientists renamed the Lowell canal the Valles Marineris.

Everything about the Valles Marineris is breathtaking. It literally spans one-fifth the circumference of Mars. At its widest, rim-to-rim, it is over 600 km (370 m) across. Like Olympus Mons, Valles Marineris is best seen in its entirety from orbit. Standing on one rim, a future explorer would not be able to see the opposite rim, due to Mars's curvature. In some places, the canyon floor is 5.7 km (3.5 m) below the mean elevation of Mars, making it one of the lowest regions of Mars.

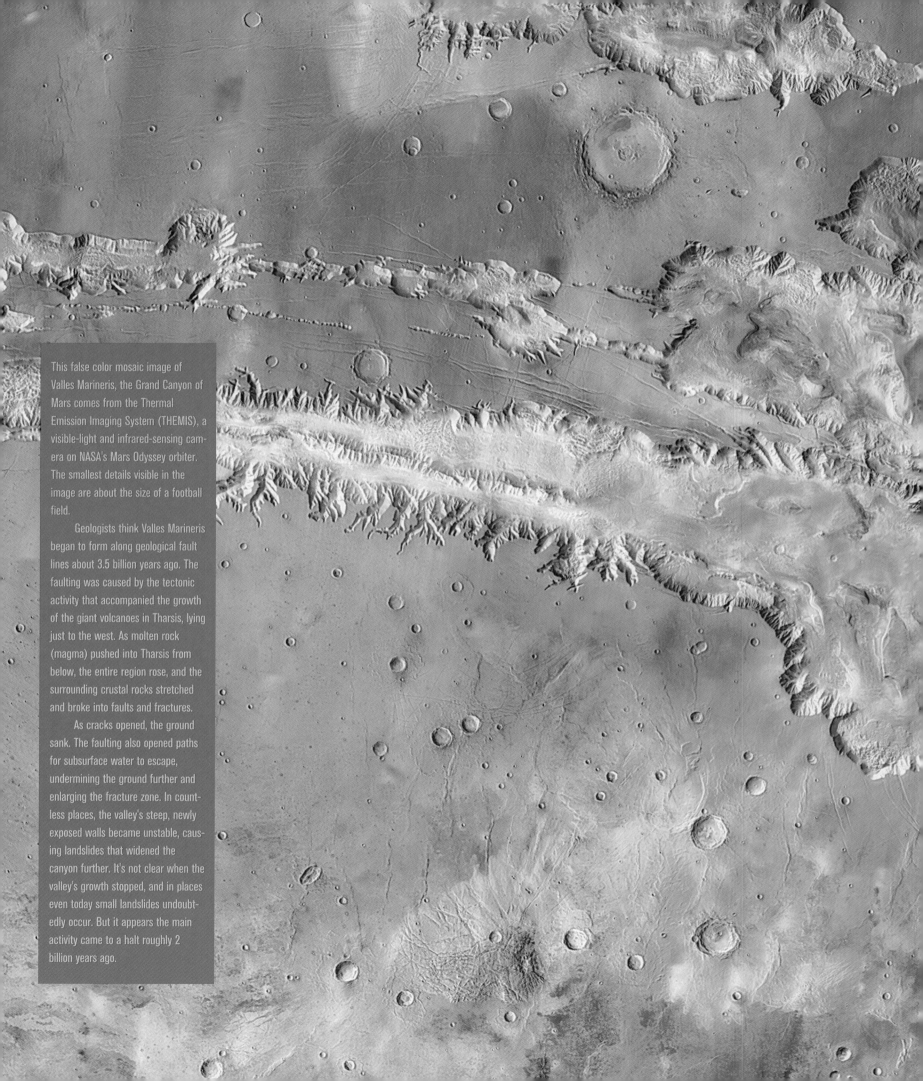

This false color mosaic image of Valles Marineris, the Grand Canyon of Mars comes from the Thermal Emission Imaging System (THEMIS), a visible-light and infrared-sensing camera on NASA's Mars Odyssey orbiter. The smallest details visible in the image are about the size of a football field.

Geologists think Valles Marineris began to form along geological fault lines about 3.5 billion years ago. The faulting was caused by the tectonic activity that accompanied the growth of the giant volcanoes in Tharsis, lying just to the west. As molten rock (magma) pushed into Tharsis from below, the entire region rose, and the surrounding crustal rocks stretched and broke into faults and fractures.

As cracks opened, the ground sank. The faulting also opened paths for subsurface water to escape, undermining the ground further and enlarging the fracture zone. In countless places, the valley's steep, newly exposed walls became unstable, causing landslides that widened the canyon further. It's not clear when the valley's growth stopped, and in places even today small landslides undoubtedly occur. But it appears the main activity came to a halt roughly 2 billion years ago.

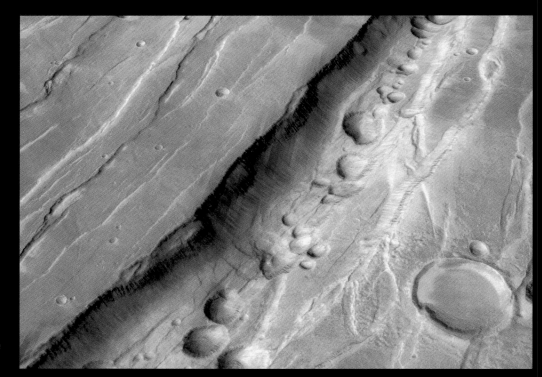

This perspective, looking down and to the north, was taken by the High Resolution Stereo Camera on board ESA's Mars Express spacecraft. It shows pits and tectonic grabens in the Phlegethon Catena region of Mars.

Because Mars is currently a desert world, scientists have had to look to a range of geological processes to explain the canyon. The Grand Canyon is clearly a water-cut feature. The Colorado River has been cutting through the Colorado plateau for millions of years. Just its size alone indicates that Valles Marineris is a far older and far more complex feature. The canyon appears to be the result of massive fracturing, landslides, and surface collapses, and water and wind action.

With newer generations of orbital spacecraft sporting greater image resolution, scientists have observed layer upon layer of exposed flat rock strata decorating the canyon walls. Still closer views, and instruments such as spectrometers, which analyze reflected wavelengths, have permitted some good guessing. Not surprisingly, because the canyon cuts across the largest volcanic dome in the Solar System, the canyon wall and floor strata appear to consist of thick layers of basalt lava flows interspersed with ash falls or fragmented lava.

The history of Valles Marineris appears to date back almost to the end of the great Noachim bombardment, about 3.5 billion years ago. The growth of the Tharsis dome and the giant volcanoes certainly exerted great stress on the Martian crust. Magma rose into the dome and pressed upward, while lava flows piled on top and pressed downward. Faults began to open in the surface rock and radiate outward across the dome. Steep walls of rock moved apart. Large blocks of rock called grabens, which run axially in the center of the cracks, dropped downward to form the canyon floor. In many places along the canyon floor, chains of pits are observed where undermined material has collapsed. Melting of subsurface ice, poorly consolidated rock, ancient lava conduits, or any combination of the above are likely to have created the chains.

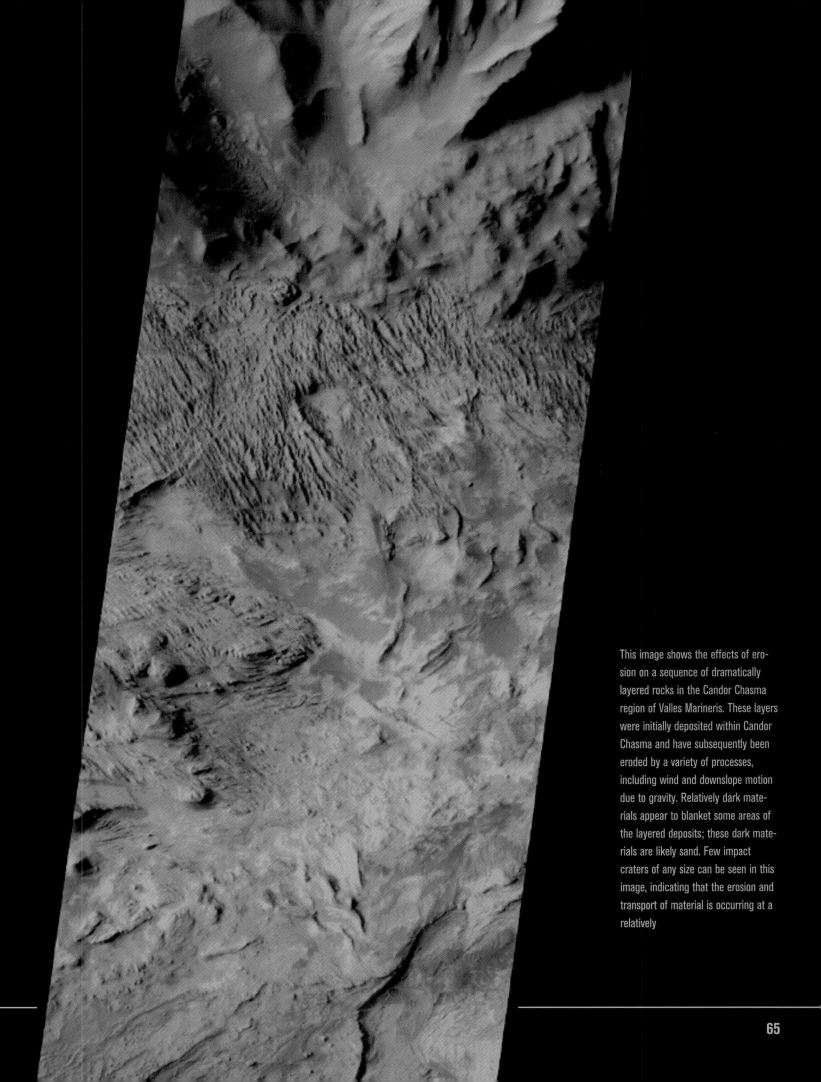

This image shows the effects of erosion on a sequence of dramatically layered rocks in the Candor Chasma region of Valles Marineris. These layers were initially deposited within Candor Chasma and have subsequently been eroded by a variety of processes, including wind and downslope motion due to gravity. Relatively dark materials appear to blanket some areas of the layered deposits; these dark materials are likely sand. Few impact craters of any size can be seen in this image, indicating that the erosion and transport of material is occurring at a relatively

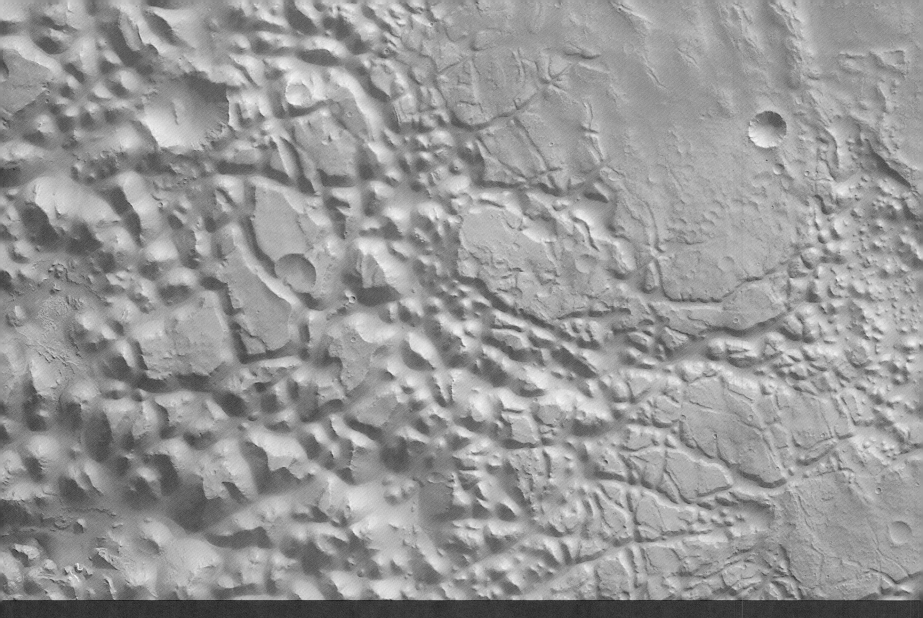

Iani Chaos is one of many regions east of Valles Marineris characterized by disrupted or chaotic terrain. The morphology of this terrain is dominated by large-scale remnant massifs, which are large relief masses that have been moved and weathered as a block. These are randomly oriented and heavily eroded. To the south (left) in this image, these mesas, which appear as flat-topped hills, range from less than 1 km to roughly 8 km wide, with a maximum relative elevation of approximately 1,000 m.

Instabilities in the steep walls, perhaps due to the melting of subsurface ice, has led to massive rim collapses in many locations along the length of the canyon. These collapses are a primary method of canyon widening and continue today. One dramatic region of wall collapse is the Candor Chasma, located on the north side almost midway along the canyon. Candor Chasma is a 300-km (180-m)-wide amphitheater-shaped indentation in the north rim of the canyon. The floor of the chasma consists of hummocky landslide deposits (rounded or conical mounds) that have flowed outward from the base of the cliff.

As the canyon continued to develop, parallel east-west faults gradually merged with the main system. Collapses broke down intervening rims and connected the parallel canyons into a giant rift system.

Water appears to have had a significant effect on the canyon's development. Striations in the collapse debris appear to indicate an outward flow across the canyon floor. Any surface or subsurface water present in the canyon's early days would certainly have drained into the cracks to further undermine the fracture zone. Later, during the Hesperian era (3.5 to 2.5 or 2 billion years ago), Mars's climate was undergoing a lengthy transition from wet to today's dry conditions. At the northeastern end, there is strong evidence of massive Hesperian era flooding that spread into the Chryse Planitia. What remains of the northeastern canyon floor is a chaotic terrain of channel scars, mesas, and hills.

In spite if its great size and abundance of geological evidence, much is still unknown about Valles Marineris. One of the important questions about the canyon relates to the significance of water. The main canyon appears to be primarily tectonic in origin, but the tributary canyons along the sides clearly exhibit the dendritic patterns that are common to water erosion features on horizontal rock layers on Earth. Are they water features, dry mass transport, or some combination? It is likely that the answer to this mystery and many others will be the domain of future Martian explorers. However, it is certain that tectonic forces, earthquakes and other crustal pressures, mass collapses, water, and wind action have all played greater or lesser roles in the creation of the canyon.

Nanedi Valles is a roughly 800-km valley extending southwest-northeast and lying in the region of Xanthe Terra, southwest of Chryse Planitia. In this view, Nanedi Valles is seen to be relatively flat-floored and steep-sloped, and exhibits meanders and a merging of two branches in the north. The valley's origins remain unclear, with scientists debating whether erosion caused by groundwater outflow, flow of liquid beneath an ice cover, or collapse of the surface in association with liquid flow is responsible.

Shown here are the regions of Granicus Valles and Tinjar Valles in the Valles Marineris. They may have been formed partly through the action of subsurface water, due to a process known as sapping.

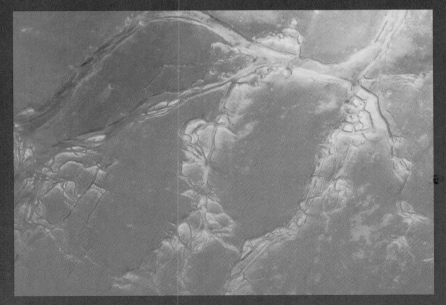

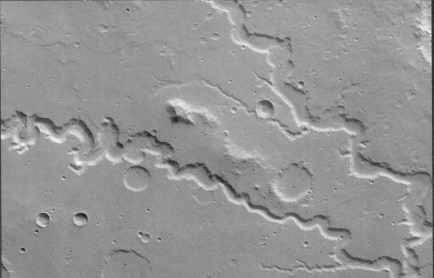

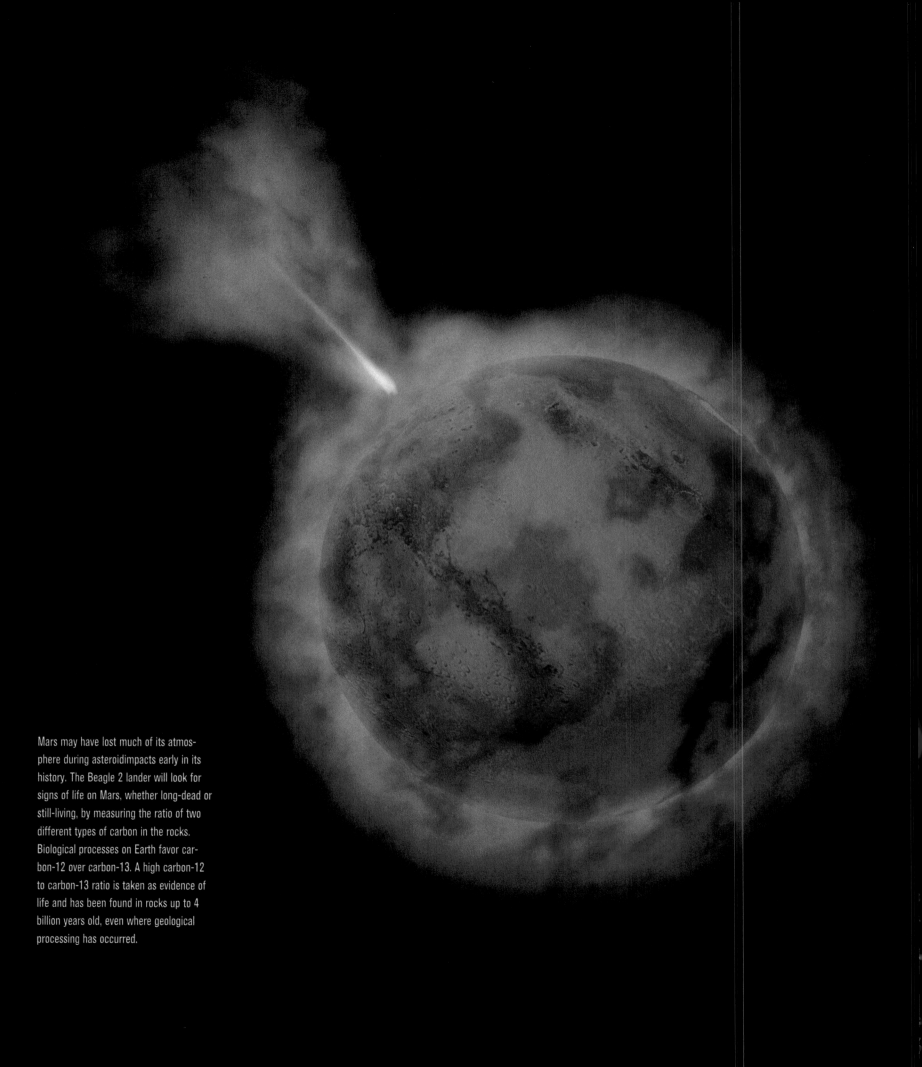

Mars may have lost much of its atmosphere during asteroidimpacts early in its history. The Beagle 2 lander will look for signs of life on Mars, whether long-dead or still-living, by measuring the ratio of two different types of carbon in the rocks. Biological processes on Earth favor carbon-12 over carbon-13. A high carbon-12 to carbon-13 ratio is taken as evidence of life and has been found in rocks up to 4 billion years old, even where geological processing has occurred.

5 MAKING AN IMPACT ON MARS

The impact of a large meteorite or asteroid on a planetary surface is among the fastest and most violent of all geologic processes. Towering mountains and deep chasms can take tens if not hundreds of millions of years to create. An impact is geologically almost instantaneous. In a split second, a rock from space screams through the atmosphere at speeds of 10 to 20 km (6 to 12 m) per second.

Ranging anywhere from a small pebble to asteroids tens of kilometers in diameter, the space rock transfers tremendous amounts of kinetic energy to the impact site. Traveling far faster than the vibrations it creates at the point of impact, the space rock punches into the subsurface before detonation takes place. In moments, a fireball of superheated gas, glowing melted rock, and dust erupts. The force is great enough to peel back the surface rock. Rock petals bend outward and form a raised rim around a bowl-shaped crater that is many times the diameter of the object that created it. A steady rain of ejecta begins. The heaviest particles fall back next to the crater, piling on top of the rock petals and further building the steeply sloped rim fanning outward. Finer ejecta materials linger in the air longer and are carried further away. They settle in a blanket of dust that may point in one dominant direction due to persistent wind effects at the time of the impact.

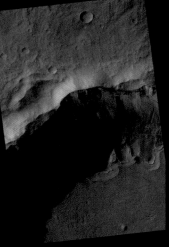

Gullies on Mars exhibit a range of structures. The large gully in the center of this image is deeply cut into with a wide alcove. The gullies on the west rim of the crater have small alcoves and tiny channels. Many of the channels appear to start at one of the fine layers along this wall. Water may have come from the underground and traveled along these layers to form the gullies by erosion. Gullies on Mars exhibit a range of structures. The large gully in the center of this image is deeply cut into with a wide alcove. The gullies on the west rim of the crater have small alcoves and tiny channels. Many of the channels appear to start at one of the fine layers along this wall. Water may have come from the underground and traveled along these layers to form the gullies by erosion.

The force of the impact is governed by the equation for kinetic energy. The energy released by the impact is the product of one-half the mass of the space rock times its velocity squared. This equation explains why small space rocks produce craters. Even small meteors, if they survive the friction of their passage through the Martian atmosphere, are traveling fast enough to pack a big wallop.

One of the most fascinating aspects of Martian impacts is that, on occasion, the force is sufficient to kick chunks of the Martian surface into space. Some of those chunks appear to have randomly crossed interplanetary space and been sucked in by Earth's gravity. A few Martian rocks have fallen to Earth as meteorites and been collected by scientists from the Antarctic ice sheets. Geochemical analysis of these meteorites demonstrates that they have a chemical makeup virtually identical to the chemical analysis of the Martian surface conducted by spacecraft. Dating measurements, though somewhat uncertain because of potential contamination, estimate the meteorite ages of somewhere between 170 million and 4 billion years. The younger meteorites indicate geologically recent volcanic activity on Mars.

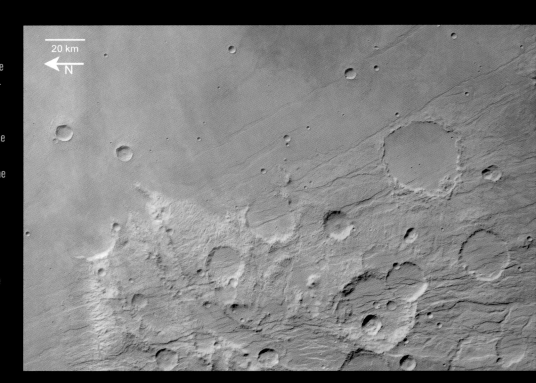

This image was taken by a camera from Mars Express. The displayed region is the eastern part of Claritas Fossae and the western part of Solis Planum. The diffuse blue-white streaks in the northern parts of the scene are clouds or aerosol sprays.

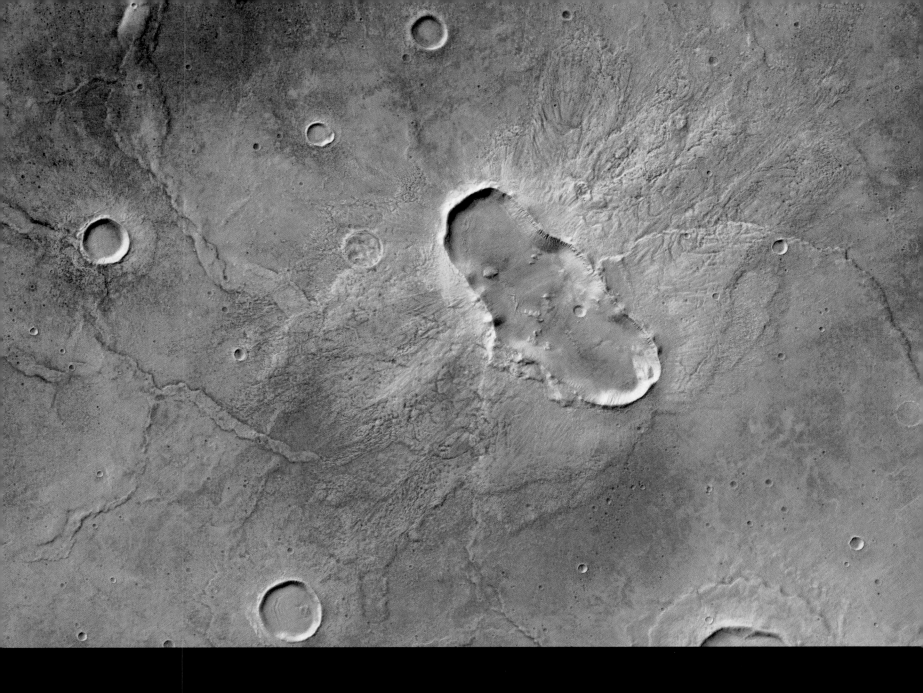

Beneath the impact site, shock waves penetrate the subsurface rock in a shatter cone fanning outward with depth. If enough heat is generated, melting occurs, and subsurface rock can slosh back to the basin's center and freeze as a low central peak. If subsurface water or ice is present, it will form steam that adds to the explosive force generated by the impact. Subsurface water will mix with the dust to form a mud-soaked slurry that spreads down a slope in toe-shaped blobs. Such impact sites are called rampart craters, and they provide important data about the presence of subsurface ice. The depth of craters can be estimated accurately with spacecraft altimetry. The presence of the rampart features means that an ice layer was penetrated. Therefore, the depth of the crater is an indicator of how deep the ice layer is in the area of the impact.

The HRSC on board ESA's Mars Express spacecraft obtained this color image of a region of Hesperia Planum. A large elliptical impact crater, called the Butterfly Crater, is visible below the surrounding plains.

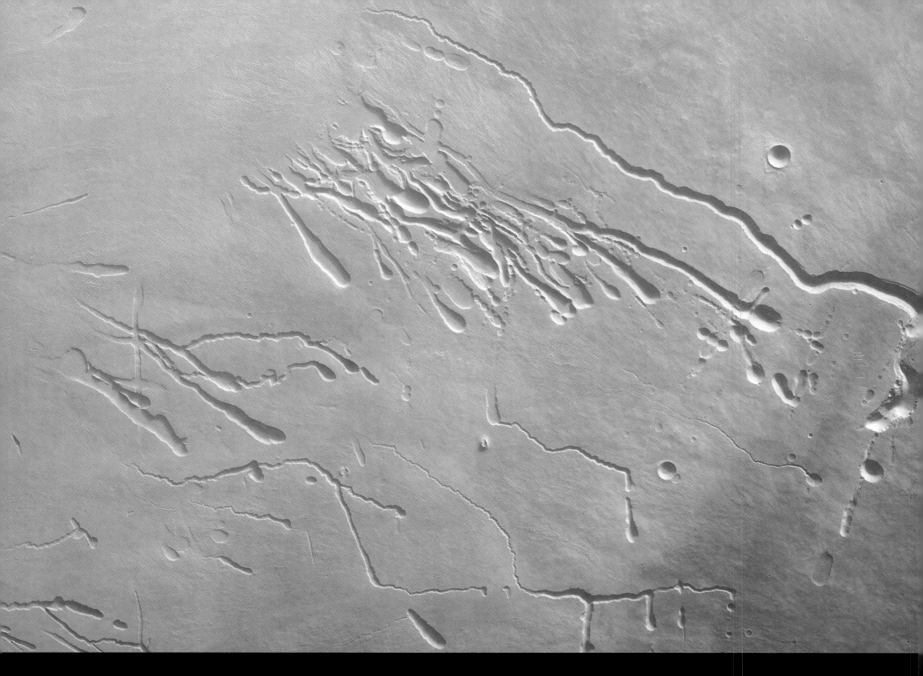

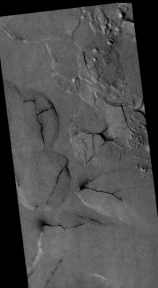

This image shows Pavonis Mons, the central volcano of the three shield volcanoes that comprise Tharsis Montes. Gently sloping shield volcanoes are shaped like a flattened dome and are built almost exclusively of lava flows. Pavonis Mons rises roughly 12 kilometers above the surrounding plains. The dramatic features visible in this image are on the southwestern flank of the volcano. It is believed that these are lava tubes, channels originally formed by hot, flowing lava that formed a crust as the surface cooled. When the lava flow stopped, the tunnels emptied and the surface collapsed, forming long depressions. Similar tubes are well known on Earth and the Moon.

This HiRISE image shows fractured mounds on the southern edge of Elysium Planitia. The mounds are typically a few kilometers in diameter and about 60 meters tall. The fractures that crisscross their surfaces suggest that the mounds, probably composed of solidified lava, formed by being pushed up from below. Smooth lava plains, which fill the low-lying areas between the mounds, are riddled with sinuous pressure ridges. The entire area is covered by a relatively thin layer of dust and sand.

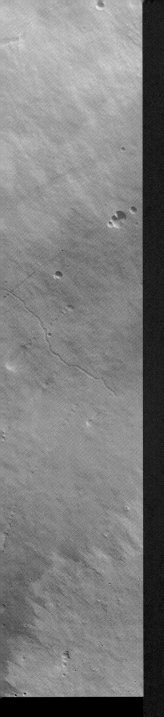

Counting Craters

Without the ability to make direct observations and to take samples, planetary geologists content themselves with estimating the ages of the different Martian surfaces by studying crater density. The technique depends upon an important assumption: the rain of meteors and other impacting debris over time is uniform. However, uniform for planetary impact studies means a declining rate, with intensive bombardment in a planet's early days and then a steady, diminishing rate. The idea is simple. The more craters, the older the surface. Generally, a Martian surface packed with touching and overlapping craters is 3.5 billion years old or older. Between 3.5 billion and 1 billion years surfaces tend to be mixed, with open spaces and scattered craters. Finally, surfaces nearly devoid of craters are likely to be less than 500 million years old.

This HRSC image provides a perspective view of residual water ice on the floor of Vastitas Borealis Crater on Mars.

The Mars Global Surveyor Mars Orbiter Camera (MOC) monitors the seasonal comings and goings of polar frost on Mars. These four wide-angle pictures of craters in both the northern and southern middle and polar latitudes show examples of frost monitoring. It is spring in the northern hemisphere, and the frost has been retreating since May. Examples of frost-rimmed craters include Lomonosov (top, left) and an unnamed crater farther north (top, right). It is fall in the southern hemisphere, and frost was seen as early as August in some craters, such as Barnard (bottom, left); later the frost line moved farther north, and we began to see frost in Lowell Crater (bottom, right) in mid-October.

The effects of an impact are greater than just the hole left behind. On Earth, an impact 65 million years ago in the Yucatan Peninsula of Mexico was large enough to affect Earth's heat balance and may have been a major contributor to the demise of the dinosaurs. Impacts on Mars would have released large amounts of dust and water into the atmosphere that would have returned in torrents and led to flooding in its early history (Noachin era).

Like all other planetary bodies in the Solar System, Mars experienced a fierce meteor and asteroid rain during the early Noachian era (4.5 to 3.5 billion years ago). The violence of the era is clearly seen in the Martian southern hemisphere in densely packed craters, craters overlapping craters, and large craters obliterating smaller ones. Large regions look Moon-like. The largest known impact in the Solar System took place in the southern Martian hemisphere. The crater left behind by the Hellas impact is in a category by itself. The structure that formed is called an impact basin and is large enough to hold an ocean.

The Martian northern hemisphere tells a different story. It has far fewer craters and looks almost as though large expanses have been wiped clean of craters. On Earth, "wiped clean" is a good descriptor of what happens to craters. Water and wind are pretty good at obliterating craters by wearing them away or filling them in. However, erosion on Mars is far less energetic, and craters remain sharply defined for millions of years. Even the oldest (billions of years old) craters on Mars, though massively worn or filled with sediment, are still visible. Why, then, are there large areas on Mars nearly devoid of craters?

Detailed orbital analyses of the northern plains of Mars—Elysium Planitia, Utopia Planitia, Amazonia Planitia, and others—reveal relatively smooth surfaces laced with fine fractures that resemble the surface of a China plate that has been broken into many pieces and glued back together. Recognizing similar features on the volcanic plains in Iceland, geologists are pretty certain that the most recent era on Mars, the Amazonia (2.5 or 2 billion years ago to the present), has been characterized by fluid basaltic lava flows that overlapped and covered much of the Martian northern hemisphere. The northern hemisphere was likely pounded as densely as anywhere else on Mars, but those craters are now mostly buried under lava flows. The few craters visible in the northern plains are recent, occurring following the lava flows.

Which Craters Came First?

Martian geologists can be pretty certain which craters are older and which are younger using two "CSI" (Crater Site Investigation) techniques. The first is erosion. Water (in ancient times), persistent wind, diurnal heating and cooling, and gravity sculpt the surface of Mars. When a meteorite slams into Mars, it blasts out a crater, depresses the surface, mounds up debris around its rim, and splatters remaining debris beyond the rim. The crater and rim have edges that appear sharp in pictures taken from orbiting spacecraft. Over time, the edges become blurred as material breaks off and settles into the crater or is carried away by winds. The crater takes on a ragged appearance as it ages. The most ancient of discernable craters appear as faint circles in an otherwise smooth plain. The second technique is like looking at a stack of books. You are fairly safe in assuming the bottom book was placed there first and the top book last. When impacts overlap, it is pretty easy to tell which impact came first. A small crater in the middle of a big crater came second. If it had been first, the bigger impact would have obliterated it. A crater that cuts into the rim of another crater is younger.

6 WATER AND WIND

The surface of a planet is like an artist's canvas. Layer upon layer of paint is applied with a brush and sculpted with a pallet knife until the picture emerges. The picture of Mars has been emerging for 4.5 billion years. It is a complex canvas built up of successive layers of lava and broad sweeps of sediment. Water, wind, and impacts have also sculpted its surface. It is a work in progress.

Although the impact of a space rock is quick and devastating, the real work of shaping the Martian surface is done by water and wind. Water and wind are gradual but immensely powerful agents of change.

In spite of its desert appearance, Mars is very much a wet planet. Currently, much of its water is locked beneath the surface in massive polar ice deposits, subsurface ice layers, and in hydrated minerals that include water in their molecular structure. It wasn't always so. During the Noachim period, the earliest age of Mars, water moved freely on the surface. Mars may have had a greater water to total-planetary-mass ratio than present-day Earth. Mars had enough surface water to form rivers, lakes, and shallow oceans. If Mars had had a uniformly level surface, an ocean 10 m (33 f) deep would have completely covered the planet.

This image, taken by the HRSC on board ESA's Mars Express spacecraft, shows a close-up view of a crater in the southern highlands of Mars, in an area called Promethei Terra, which is east of the Hellas Planitia impact basin. The smooth surface is caused by a layer of dust or volcanic ash that is up to several tens of meters thick. The numerous dark lines to the northwest of this crater are 'dust devil' tracks.

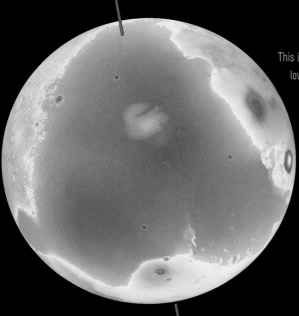

This is Mars as it might have appeared more than 2 billion years ago, with an ocean filling the lowland basin that now occupies the north polar region. Topographic deformation of features that ring the basin, which are thought to be shorelines formed by the ancient ocean, suggests that Mars experienced reorientation of the planet relative to its rotation axis. The margins of the ocean shown here account for the topographic deformation that would have resulted from this reorientation. Sinuous features near the top of the image are valleys carved by large floods that may have supplied the ocean water. The image was generated using Viking Orbiter images and topographic data from the Mars Orbiter Laser Altimeter on board the Mars Global Surveyor spacecraft.
Credit: Taylor Perron/UC Berkeley

For surface water to exist on Mars, the early Martian atmosphere needed to be much denser than it is today. Spacecraft measurements of isotopes of Martian atmospheric gases indicate that as much as 90 percent of the initial Martian atmosphere has since escaped into space. The pressure of the early atmosphere would have been sufficient for liquid water to exist on its surface and not boil away, as it would today.

Supporting the concept of a watery surface on early Mars are many surface markings. One water signpost is found in some of the craters of the southern hemisphere. The southern hemisphere exhibits some of the oldest terrain on Mars. One would expect that ancient impact craters would show signs of weathering, but weathering for some craters has taken a dramatic twist. The craters look soft. Their sharp initial rims have been smoothed, as though they had melted and partially filled in their crater floor. Fresh craters have sharp rims that gradually slope, bowl-like, toward their centers. The rims of these craters are more like low concentric circular ridges that separate the surrounding plain from the plain of the crater floor. The effect is called terrain softening, and geologists were hard pressed to explain "softened craters" on a desert planet when they were first discovered. Recognizing that there are large amounts of water on Mars has provided insights into the softening process. On Earth, water-saturated steep slopes often become unstable and slumping occurs. From above, slope failure produces a softened appearance similar to the softened terrain of Mars. Water adds mass to soil and lubricates soil movements. In areas of frozen soil on Mars, the heat generated by an impact would produce melting, and released water might trigger slope movements.

This HRSC image shows a system of "sapping channels" in Louros Valles (named after a river in Greece), south of the Ius Chasma canyon. Sapping is erosion by water that comes from the ground as a spring or seeps from between layers of rock in a wall of a cliff, crater, or other type of depression. The channel forms from water and debris running down the slope from the seepage area. The Ius Chasma belongs to the giant Valles Marineris canyon system. The Geryon Montes, visible at the northern border, is a mountain range that divides the Ius Chasma into two parallel trenches. The dark deposits at the bottom of the Ius Chasma are possibly the result of water and wind erosion.

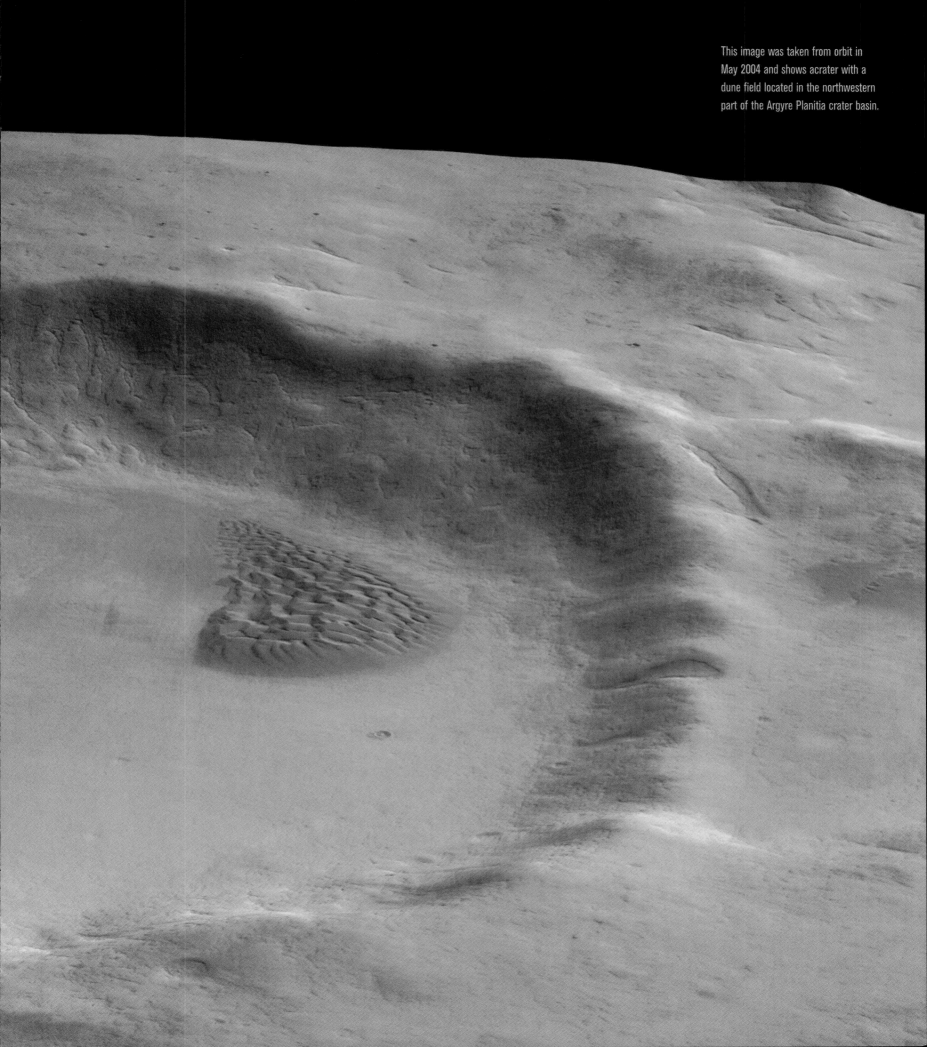

This image was taken from orbit in May 2004 and shows acrater with a dune field located in the northwestern part of the Argyre Planitia crater basin.

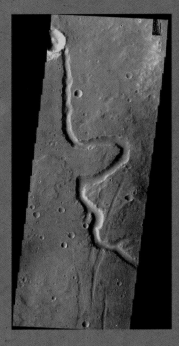

Another important observation of the watery nature of Mars is the presence of braided and meandering channels on the Martian surface. The channels are dry today, but they look identical to water-cut channels found on Earth, especially in desert regions, where water is intermittent. Ancient channels stretching hundreds of kilometers in length are clearly seen entering the southern hemisphere's Hellas Planitia from the surrounding mountainous regions. The floor of the impact area is layered, indicating deposition of loosely consolidated gravels and dust deposited by water and wind.

Another giant impact site of the southern hemisphere is the Argrye Planitia. This appears to have been a temporary stopping point in what became a global scale river system. Meltwater from the south polar cap found its way into the impact basin and turned the floor into a large circular lake. Later, water, and possibly wind, broke through the northern rim of the impact crater, permitting water to drain northward. A massive waterway was carved that ultimately ended in the Chryse Planitia 10° north of the equator. The channels entered Chryse just south and to the east of another channel system pointing eastward off the Tharsis volcanic bulge.

This image was acquired by Opportunity using its panoramic camera. It shows a large wind-blown ripple called "Scylla" and other nearby ripples and patches of brighter rock strewn with dark cobbles. The scours and ripple crests are probably due to the presence of basaltic sands mixed with hematite-rich spherules. Patterns on the larger ripple flanks were caused by different amounts of reddish dust. The larger ripple flanks have an intricate mixture of erosional scours and secondary ripples extending downward from the main ripple crests, suggesting that these ripples have most recently encountered a period of wind erosion and transport of their outer layers.
Image Credit: NASA/JPL-Caltech/Cornell

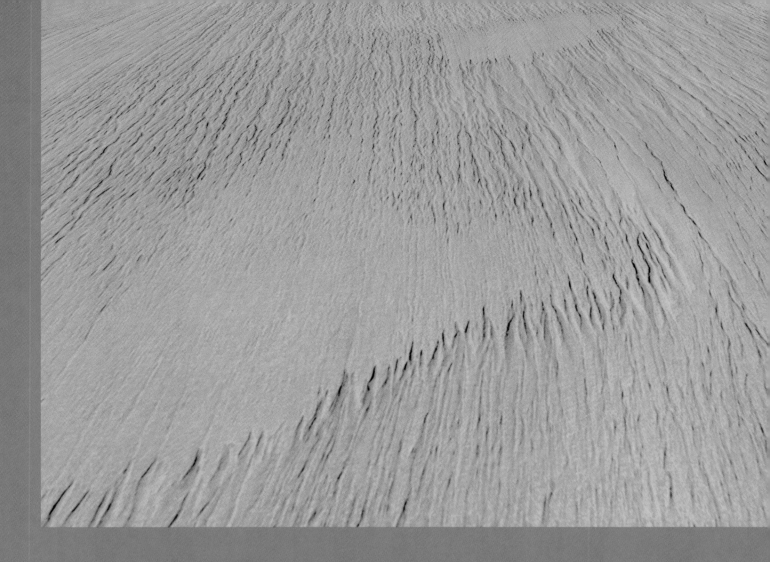

Northward drainage with resulting sediment deposition is partially responsible for the relatively smooth and somewhat featureless appearance of large stretches of the Martian northern hemisphere. However, water isn't the only story. Wind plays a role, too.

In spite of Mars's very thin atmosphere, about one one-hundredth the density of Earth's, strong winds can kick up some massive dust storms. (Recall the planet-wide storm that obscured Mars in 1971.) Wind-sculpted features abound across Mars.

The most obvious Martian wind-created surface features are sand and dust dunes. The dunes are exceptionally easy to see and identify. They look just like the dunes of the Sahara and Mojave Deserts on Earth. Large fields of barchan dunes speckle broad plains. A barchan is a curved, smile-shaped (when looked at from the right direction) dune. The outer or upwind side is a smooth convex curve that is symmetrical to the direction of the wind. The lee side of the dune has a concave curve with a slip face, where mini sand avalanches occur. The ends of the dune form two "horns" that point in the direction the wind is blowing. Slowly, barchan dunes form across the Martian surface as sand from the upwind side is blown up and over the slip face. Wind eats away one side of these dunes and builds up the other.

The images of yardangs—features sculpted by wind-blown sand—seen here south of Olympus Mons on Mars were obtained by the HRSC on board the ESA Mars Express spacecraft.

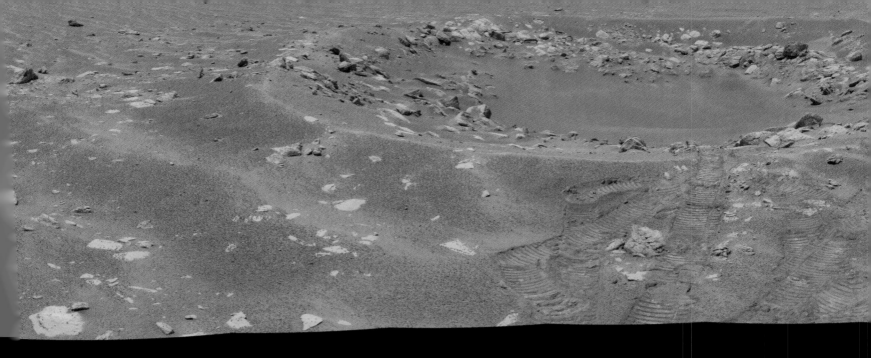

This view, in approximately true color, reveals the structure of an impact crater informally named "Fram," located in the Meridian Planum region of Mars. The picture is a mosaic of frames taken by the camera on NASA's Mars exploration rover Opportunity in April 2004. The crater spans about 8 meters (26 feet) in diameter. Opportunity paused beside it while traveling from its landing site toward a larger crater farther east. This view combines images taken using three of the camera's filters for different wavelengths of light.

Image Credit: NASA/JPL/Cornell

Water Below

Radar echo sounding, a technique used on Earth to probe the insides of glaciers, has been exported to Mars. The European Mars Express spacecraft carries an instrument called the Mars Advanced Radar for Subsonic and Ionospheric Sounding, or MARSIS. MARSIS has probed the Martian south pole and mapped a layer of ice below the surface that was first identified in the 1970s. The layer, about 4 km (2.5 m) below the surface, is massive enough to create a planet-wide ocean, were it to melt, 11 m (36 f) deep (assuming Mars had a uniform surface).

The water mass under the pole is still not sufficient to explain the extent of water erosion features seen across the planet. The amount of water locked in the polar reservoir is only about one-tenth to one one-hundredth of the amount of water that once flowed on the planet. Are there other reservoirs, or has most of the Martian water been lost to space?

NASA's Mars exploration rover Spirit used its navigation camera to record this scene on the day the rover arrived at the crest area of "Husband Hill" inside Gusev Crater. The rover had just completed its longest drive in months, 44.8 meters (47 feet), before taking this picture. A wind-sculpted ripple of sand or dust dominates the fore-ground, which is the top of the hill, while a whirlwind lofts a column of dust above the plain in the distance.
Image Credit: NASA/JPL

Other familiar dune shapes on Mars are linear dunes that form long parallel ridges perpendicular to the wind direction. These dunes are often slightly sinuous in shape. Linear features also form around resistant objects, such as crater rims. Wind parts around these objects and moves sand and dust to form long trailing deposits that merge to form a cone-shaped wind shadow downwind.

Dust devils are yet another wind erosion discovery on Mars. Scientists, observing pictures from the Spirit and Opportunity spacecraft, were startled to see dust devils cross the field of view of the rover cameras. Like trains of mini tornadoes sometimes seen on Earth, lines of dust devils cross the broad plains. Quickly rising warm air on Mars triggers the formation of dust devils. Surrounding air rushes in to fill the void, and spiraling begins. The discovery of dust devils enabled scientists to explain some weird worm-like dark tracks on some otherwise light-colored Martian plains—dust devil tracks.

Like Earth dust devils, Mars dust devils may be electrically charged. This could lead to problems with dusty "static cling" on the suits of future Martian explorers.

NORTH AND SOUTH

This image of the Martian north polar ice cap shows layers of water ice and dust in perspective. Here we see cliffs, which are almost 2 kilometers high. The dark material in the caldera-like structures and dune fields could be volcanic ash.
Image Credit: ESA/DLR/FU Berlin (G. Neukum)

That Mars has bright patches surrounding its rotational north and south poles and the patches change in size with the seasons is something that has been known for a long time. William Herschel, discoverer of the planet Uranus, proclaimed the bright patches on Mars to be arctic-like ice fields. Consequently, it was first assumed that the poles were covered with layers of water ice that melted in the summer to provide moisture to the lower latitudes. Later, with the realization that the Martian atmosphere is predominantly carbon dioxide, it was suggested that the poles were actually covered with dry ice.

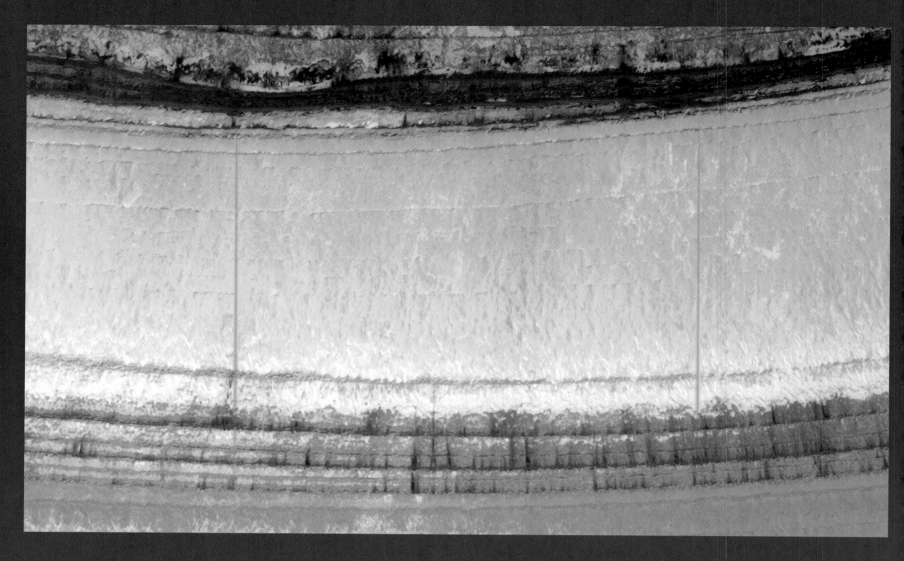

This image shows layering within the northern hemisphere's ice cap, which probably reflects seasonal variability in accumulation of ice versus sublimation (going from a solid to a gas). The presence of sand dunes indicates the transporting of sedimentary materials by wind. Erosion is seen as a series of undulating ridges between the layered terrain and the dune field. Near the top of the image several vents occur where materials from the shallow subsurface have erupted onto the surface. Credit: NASA/JPL/University of Arizona

As it turns out, the Martian poles have both. Each has a centrally located permanent water ice core layer surrounded and covered over by a larger dry ice deposit that comes and goes with the seasons. It is the dry ice fields that were observed to change in size. With the coming of summer in the Martian south pole, the dry ice polar coating sublimates directly into gas and adds to the overall quantity of carbon dioxide in the atmosphere. (Carbon dioxide freezes at -79° C or -110° F and boils at -78.5° C or -109° F.) At the same time, the Martian north pole cools, and winter in the north begins. Carbon dioxide gas condenses into dry ice and expands the extent of the northern polar cap.

A year later, the Martian seasons reverse, and the south pole cap recedes while the north pole cap grows. The water ice cores of both poles remain fairly constant in size, although some sublimation of water into the atmosphere occurs at the north pole during its summer. Sublimation of water doesn't appear to be taking place at the south pole in spite of the fact that summer in the south tends to be warmer than summer in the north. One possible explanation for the difference is that the warmer southern summer encourages the formation of dust devils that loft dust particles into the southern atmosphere. The dusty atmosphere diminishes the warming effect of sunlight on the polar ice cap itself, leading to slightly cooler summer temperatures near the pole.

Sand-laden jets shoot into the polar sky in this view by noted space artist Ron Miller. It shows the Martian south polar ice cap as spring in the southern hemisphere begins.
Image Credit: Arizona State University/Ron Miller

This Mars Orbiter Camera (MOC) image shows south polar mesas composed largely of solid carbon dioxide separated by circular depressions. The curved scarps, which form the boundaries of the mesas, retreat approximately 3 meters per Mars year (two Earth years) owing to sublimation, which occurs primarily during the Martian summer months.
Image Credit: NASA/JPL/Malin Space Science Systems

Both poles rest on elevated mounds that rise about 3 km (2 m) above the surrounding plains. With the summertime clearance of dry ice frost, the mounds reveal concentric parallel lines ringing the poles in irregular circles. The lines are the horizontal edges of layer upon layer of deposits of sedimentary rock, with each layer slightly more weathered away and smaller than the one below. Scientists have concluded that the poles are major areas of sedimentary deposition. The amount of deposition appears to change over the long term and is possibly linked to changes in Mars's axial tilt. The layers are probably made from a combination of sand, dust, and ice.

Spring is a violent time at Mars's south pole. It starts innocently, though. Through the intensely cold Martian winter, carbon dioxide gas condenses and freezes, gradually forming a meter-thick layer of dry ice. As spring and summer approach, sunlight passes through the ice to warm the dark underlying rock. The dry ice, in contact with the warm rock, begins to sublimate directly into gas. The remaining surface ice layer forms a seal and creates a gas pocket. For perhaps 60 days or more, all is quiet as pressure slowly builds. Then, the dry ice layer starts lifting.

If you were an astronaut walking across the ice at this time, it would be a good idea to pick up your pace! Suddenly, a weak spot in the ice layer ruptures. High-pressure gas roars through the opening at speeds of up to 150 km (90 m) per hour. Chunks and snowflakes of dry ice fly upward and spread out in small fans. Dark sand, scooped up by the racing gas, forms long fan-shaped markings on the white dry ice surface. It was these kinds of markings, growing larger and darker over time, that were first spotted in satellite pictures. Equally mysterious were spidery grooves observed running along the bottom of the ice layer and converging at the apex of the fans. The groves appear to be another artifact of the gas release. Sand-laden gas scours and carves spindly channels along the bottom of the ice layer.

Another strange result of dry ice sublimation is an effect dubbed "Swiss cheese terrain." Anyone familiar with northern winters will understand the effect. In the springtime, snow melts unevenly, and roundish patches of dark soil peek out. The dark patches warm more rapidly than the white snow surrounding them. The patches enlarge as the snow at the edges melt. From above, snow melting in the north easily resembles the Martian Swiss cheese melting. Circular depressions form due to unequal sublimation of dry ice. Gradually, the depressions widen, as the Martian spring advances, and grow together.

The Martian poles offer many mysteries. Does the water ice move as do glaciers on Earth? There is some indication that it does around its edges. Ice seems to have divided around craters along the edges of the ice fields. Furthermore, there is a distinct absence of impact craters in the water ice fields. Either the fields are very young, or flow takes place that obliterates the craters. Ice flow would indicate very thick ice layers. Glacial ice has to accumulate to thicknesses of 50 m (150 f) or more before it starts to flow at its bottom. In the low Martian gravity, thicker ice layers would have to build up before pressures would be great enough for flow.

Another polar mystery is whether or not it snows on Mars. Presently, water snow is not likely, because the quantity of water in the atmosphere is very small, only about 0.03 percent of the total atmosphere. (For comparison, Earth's atmosphere is about 1 percent water.) In the past, the percentage would have been very different, as can be seen from the many water-erosion surface markings. Dry ice snow is another matter. Dry ice clouds are common over the Tharsis volcanoes and the poles. Scientists speculate that dry ice snow falls are quite likely over the poles during long winter nights.

The greatest mysteries of the Martian poles will have to wait until polar robot landers and human explorers arrive. Like Earth ice fields, the ice of Mars will hold important traces of atmospheric gases that can be dated like tree rings. The poles are the Martian library of volcanic events, impacts, and polar tilting waiting to be deciphered.

This 2005 picture is a composite of daily global images acquired during a previous Mars year. The picture shows the south polar region of Mars. Image Credit: NASA/JPL/Malin Space Science Systems

This is a wide angle view of the Martian north polar cap in early summer. The picture was acquired near the start of the mapping phase of the Mars Global Surveyor mission. The light-toned surfaces are residual water ice that remains through the summer season. The nearly circular band of dark material surrounding the cap consists mainly of sand dunes formed and shaped by wind. The north polar cap is roughly 1,100 kilometers across. Image Credit: NASA/JPL/Malin Space Science Systems

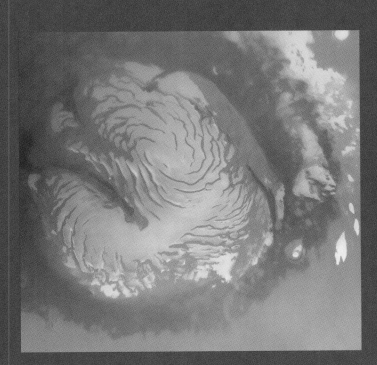

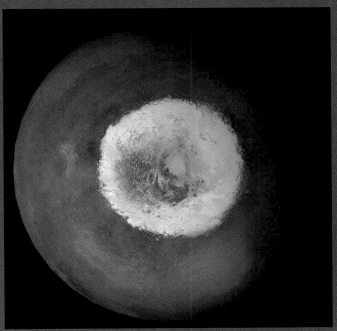

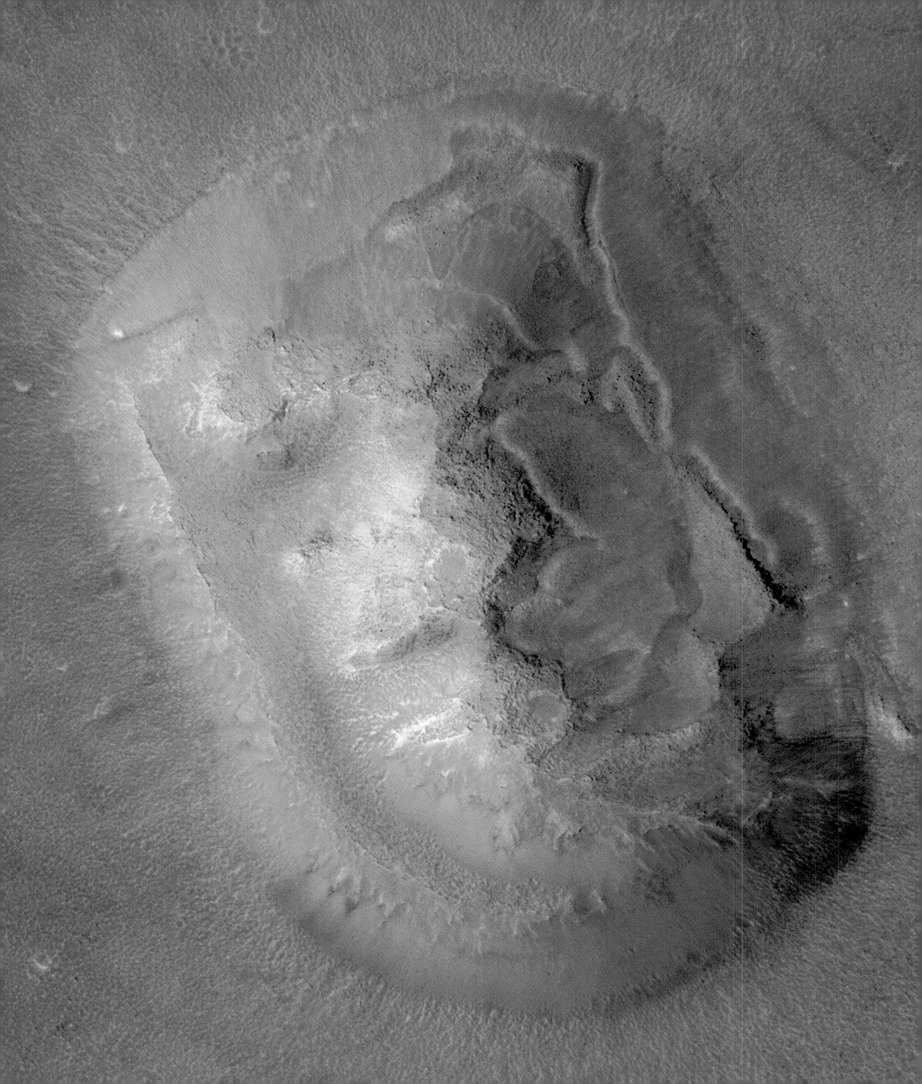

8 LIVING OR NOT?

In the mid-1950s Walt Disney had on his Sunday night broadcasts a wonderful segment called the "Tomorrowland" series, which focused on the future of space exploration. The first episode, Man In Space, was aired three years before the first satellite launch. It portrayed how humans would first go into space. Famous rocket scientists presented their ideas. The same year the first artificial satellite reached orbit (1957), the third episode, entitled Mars and Beyond, aired.

The highly creative (and possibly mildly loony) animation artists at Disney created a wonderful, though fanciful, sequence on what life on Mars might be like. They produced many fantastic images: wandering sea serpent-like plants in search of better soil; flying saucer creatures with transparent bodies that focused sunlight to simultaneously capture and cook their prey; hunter plants that puffed poison gas at flying insects; and majestic silicon-based crystal creatures that formed delicate towers during the day that shattered in the cold of the night.

This is the first photograph ever taken on the surface of the planet Mars. It was obtained by Viking 1 just minutes after the spacecraft landed successfully. We see both rocks and finely granulated material, either sand or dust. Many of the small foreground rocks are flat with angular facets. Several larger rocks exhibit irregular surfaces with pits, and the large rock at top left shows intersecting linear cracks. Extending from that rock toward the camera is a vertical linear dark band, which may be due to a one-minute partial obscuration of the landscape due to clouds or dust intervening between the Sun and the surface. There are a number of other furrows and depressions and places with fine-grained material elsewhere in the picture.
Image Credit: NASA/JPL

None of the Disney creations were serious, but their real message was that Mars is another planet, a new world. Any life found there is most likely to be totally different from anything we are familiar with on Earth. More than fifty years later, Mars scientists are still looking for signs of Martian life with the same thought in mind.

Interest in life on Mars has been intense for more than a hundred years. The "canal" discovery and seasonal color change optical illusions took a long time to explain. The "face on Mars" optical effect captured by the Viking orbiters (see box on page xx) took less time to explain, but pseudo scientist proponents continue to be resistant to replacing fantasy with data.

In spite of these misguided attempts to prove that there is life on Mars, reasoning proponents of life on Mars point to Mars's environment. Though harsh, it is not unlivable. Most scientists agree that if there is Martian life, it will probably be microorganisms thriving in the soil near ice, in liquid water layers, or in deep cracks, lava tubes, or caves. Mars life might take on forms barely recognizable to us as living. Because of that, scientists have been scouring Earth for extreme forms of Earth life that might be capable of surviving the Martian environment. Called extremophiles, these rugged life forms could offer clues as to what to look for on Mars, even in the driest, coldest, and most radiation intensive environments.

Nineteen years after Mars and Beyond, the first true Mars search-for-life science experiments were begun. Two robot spacecraft, part of the Viking 1 and Viking 2 missions, touched down on the surface of Mars. One of their primary missions was to look for signs of life. However, they were not there to chase wandering plants or to look for crystal towers. If any life existed on Mars, they might be captured by imaging cameras, but the real-life detective work was the job of small internal chemical and biological laboratories. Each about the size of a breadbox, these laboratories were designed to look for evidence of life processes at work in Martian sediments.

About Face

Mars scientists cringe every time someone asks about the "face" on Mars. It's not that any scientist wouldn't be delighted at the discovery of intelligent life on Mars. It is just that the face doesn't exist. But conspiracy theories die hard.

The Mars face conspiracy theory began with a Viking orbiter picture of the Cydonia region of Mars. Cydonia is one of the chaotic terrain areas of Mars, covered with hard volcanic cap rock possibly cut by faults and water eroded gullies. Erosion has turned the region into the classic mesa and butte topography so familiar in western movies.

In one picture, taken when the Sun was low in the sky and shadows were long, a single butte appeared to be carved into the shape of a face. Two shadowy eyes and a straight mouth formed the face, and the head was surrounded by a strange hairdo. Scientists immediately knew the face was just a combination of erosion and shadows. Earth's surface is home to many similar geologic features that look like face pro-

files, reclining bodies, and skulls. No one claimed that these were signals from some alien race. It was a different story with the Mars face.

According to some "investigative" journalists and self-professed "science experts," the face on Mars was a sure sign of alien intelligence. The more the Mars scientists tried to explain the face's true nature, the more some of the face proponents claimed "government cover-up." The controversy reached a new low with the unhappy loss of the Mars Observer spacecraft that experienced a fuel line failure, destroying the mission as it was about to go into orbit. A few vocal protestors claimed that NASA had sabotaged its own billion-dollar spacecraft to prevent the public from seeing proof that the face was real. When the Mars Global Surveyor approached Mars, a special effort was made to take better pictures of the face. With sunlight coming from a different angle, the facial features were gone and replaced by the normal bumps and crags of an eroded butte. Were the conspir-

The first Mars Orbiter Camera (MOC) flew on Mars Observer, a spacecraft that was lost in 1993, before it reached the Red Planet. Now, after fourteen years of effort, a MOC finally has been placed in mapping orbit by the Mars Global Surveyor. On the first day of the mapping phase of the MGS mission in March 1999, MOC was greeted with this view of the "Happy Face Crater" (center right) smiling back at the camera from the east side of Argyre Planitia. This crater is officially known as Galle Crater. The bluish-white tone is caused by wintertime frost.

acy theorists finally satisfied? No. They claimed now that NASA had "doctored" the images to delete the face and keep the truth from the public

The boulder-strewn field of red rocks reaches to the horizon nearly 2 miles from Viking 2 on Mars' Utopian Plain. Scientists believe the colors of the Martian surface and sky in this photo represent their true colors. Fine particles of red dust have settled on spacecraft surfaces. The salmon color of the sky is caused by dust particles suspended in the atmosphere. Color calibration charts for the cameras are mounted at three locations on the spacecraft. Note the blue star field and red stripes of the flag. The circular structure at top is the high-gain antenna, pointed toward Earth. Viking 2 landed September 3, 1976, some 4,600 miles from its twin, Viking 1, which touched down on July 20.
Image Credit: NASA/JPL

Months before the Viking missions were launched, spacecraft designers sterilized the lander craft and safely encased them in bioshields. The reason for that precaution was both simple and essential. It would do little good to land on Mars looking for life, only to discover that Earth microorganisms hitched a ride. Furthermore, the interaction of Earth life with any indigenous Mars life could be disastrous to the Martian flora and fauna.

NASA's Viking 1 orbiter spacecraft photographed this region in the northern latitudes of Mars on July 25, 1976, while search-
ing for a landing site for the Viking 2 Lander. The speckled appearance of the image is due to bit errors. Bit errors comprise
part of one of the "eyes" and "nostrils" on the eroded rock that resembles a human face near the center of the image.
Shadows in the rock formation give the illusion of a nose and mouth. Planetary geologists attribute the origin of the formation
to purely natural processes. The feature is 1.5 kilometers (1 mile) across.
Image Credit: NASA/JPL

In an amazing feat of spacecraft technology, both landers extended robot arms
that scooped up Martian sediment and deposited it into a funnel leading to test
chambers inside the laboratories. When the sediment was doused with water,
strange chemical reactions occurred, but no life processes were identified. Life on
Mars was still an open question. Scientists concluded that the wrong questions
had been asked.

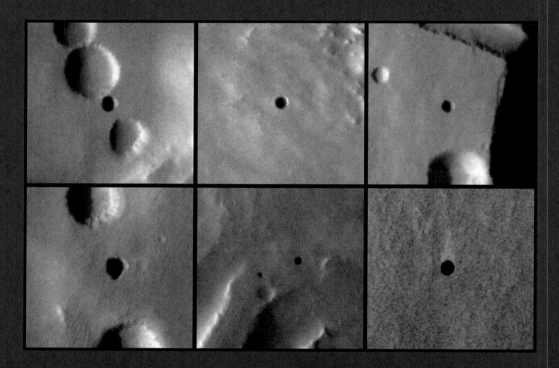

Seven very dark holes on the northern slope of a Martian volcano have been proposed as possible cave roof openings, based on day-night temperature patterns suggesting they reveal subsurface spaces. These images taken by the Thermal Emission Imaging System camera on the Mars Odyssey orbiter show the seven openings. The volcano is Arsia Mons. The features have been given informal names to aid comparative discussion. They range in diameter from about 100 meters (330 feet) to about 225 meters (740 feet). Arrows signify north and the direction of solar illumination.
Image credit: NASA/JPL-Caltech/ASU/USGS

Scientists did have a brief thrill in 1996 when a small team of NASA scientists announced the discovery of what appeared to be fossil bacteria in Martian meteorites. A small class of meteorites, found on the Antarctic ice fields, is believed to have come from Mars. Antarctica is an ideal place to search for meteorites because the dark space rocks easily stand out from the white ice fields. It is hypothesized that major impacts on Mars would have kicked Martian rock into space and some pieces would randomly find their way to Earth. The proposed Martian meteorites have the identical atomic makeup that orbital spacecraft and landers have identified on Mars. The Martian meteorites are chemically different from lunar rocks (collected by the Apollo missions) and from the usual meteorites found all around the world.

Cave Martians?

On Earth, caves are rich places for looking for exotic life forms. In them you can find blind fish, albino crickets, and slimy dripping bacteria ("snotties"), among other odd creatures. Recent images from the Mars Odyssey spacecraft indicate that Mars may also have caves. A string of football field-sized holes have been detected near a large volcano. One of the holes appears to be at least 130 m (425 f) deep. Mars caves would be ideal places to look for life. The scientists that found the holes are now dreaming of Martian spelunker robots.

Microscopic examination of a particular Martian meteorite—the ALH84001 rock, which was named for the Allan Hills location on the far western ice field, where it was collected in 1984—revealed microscopic rod shapes that could be the remains of Martian microorganisms. The 1.93-kg (4 1/4-lb) meteorite is believed to have fallen from space 13,000 years ago and is probably over 4 billion years in age.

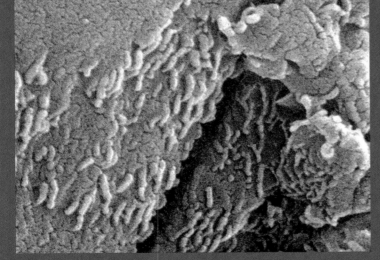

Mars meteorite, the ALH84001 rock seen through a microscope.

Naturally, the discovery caused a scientific uproar and some angry comments. Although scientists around the world would be thrilled by confirmed Martian life, acceptance of new discoveries is never easy. Using the Sherlock Holmes approach ("When you eliminate all other possibilities, whatever remains, however improbable, must be the truth.") critics hammered the team with many questions and alternate explanations. The team was able to address many of these questions but not all. Whether or not the rock contains Martian fossils is still open, and some say the chances are 50-50. Perhaps the next Martian meteorite discovery will contain more information, or the question of life on Mars will have to wait until humans set foot there. Then, for sure, the answer will be yes—Earth astronauts and the plants and microorganisms they bring with them. But maybe......

Mars Global Surveyor (MGS) operated in orbit around Mars for over nine years, longer than any other spacecraft to Mars. A few of the mission's many important discoveries include (1) Bright new deposits in two gullies on Mars, suggesting that liquid water has carried sediment down them in the past seven years. This is strong evidence that water still flows occasionally on the surface of Mars. (2) There are concentrations of a mineral that often forms under wet conditions, fine-grained hematite. This discovery led to selection of a landing site for the Rover Opportunity. (3) The fact that Mars once had a global magnetic field like Earth's, shielding the surface from deadly cosmic rays. (4) Twenty recent new impact craters. The long life of the mission also allowed MGS to track surface changes through repeated annual cycles.
Artwork Credit: Corby Waste

9 FOR A CLOSER LOOK

The history of Mars exploration is as much about technology as it is about discovery. Nearly every Mars discovery was preceded by the development and application of a new technology. To the ancients, Mars was a bright wandering star that occasionally looped backwards before continuing its usual eastward course through the stellar background. In the 1600s, the newly invented telescope permitted astronomers to resolve Mars into a disk and detect shadowy surface markings.

By the late 1800s, telescope improvements brought the Martian surface into sharper focus, although some of the observed features were wildly misinterpreted. Mars study took a giant leap in the 1960s and 70s with spacecraft that flew by, orbited, and even landed on Mars. Spacecraft permitted direct collection of data and radioed home stunning images with great detail. Today's technological advances include new and more sensitive instrumentation, better imaging systems, radar subsurface mapping, increased data storage and communication capabilities, robot surface rovers, drills, life-detection laboratories, and atomic analysis.

Because of spacecraft, we are beginning to really know Mars. Since the first attempted Mars mission in 1960, more than 40 spacecraft have aimed for the Red Planet. Unfortunately, due to launch failures, exploding fuel lines, and computer program and mathematical errors, less than half of the Mars missions have been successful. Nevertheless, the successful missions have been spectacular.

As it drew closer to Mars, the Mariner 4 spacecraft captured this image of a 151 kilometer-diameter crater in the southern hemisphere. The highly worn crater, pocked with overlying smaller impacts, has been dubbed the Mariner crater.
Photo Credit: National Space Science Data Center

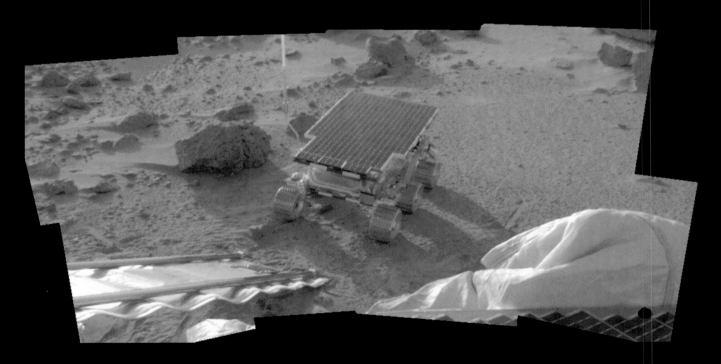

This Sol 2 mosaic from the Mars Pathfinder mission shows the newly deployed rover sitting on the Martian surface. The losslessly-compressed, multispectral insurance panorama was designed to protect against camera failure upon deployment. Had the camera failed, the insurance panorama would have been the main source of image data. However, the camera deployment was successful. Ironically, the insurance panorama contained some of the best-quality image data because of the lossless data compression and relatively dust-free state of the camera and associated lander on Sol 2.
Image Credit: NASA/JPL

This bird's-eye view combines a self-portrait of the spacecraft deck and a panoramic mosaic of the Martian surface as viewed by the rover Spirit. The rover's solar panels are still gleaming in the sunlight, having acquired only a thin veneer of dust two years after the rover landed and began exploring the Red Planet. Spirit captured this panorama on the summit of "Husband Hill" inside Mars' Gusev Crater.
Image Credit: NASA/JPL-Caltech/Cornell

The Mariner 4 mission was spectacular because it returned the first close-up images of Mars, which discovered craters in the narrow planetary slice it observed. Mariner 9 discovered the giant Valles Marineris canyon and revealed the true nature of Nix Olympica. The Viking missions provided the first surface images and chemical and meteorological data. Sister orbital spacecraft returned orbital images that vastly improved on Mariner 9 images. In 1997, the Pathfinder surface rover, named Sojourner, rolled just over 100 m (330 f) as it poked and probed its landing site during its 83-day lifespan. In 2004, two robust rovers, Spirit and Opportunity, began surface operations. Together, the two rovers have sent back more than 100,000 close-up pictures and, using spectrometers, have returned a stunning quantity of chemical and mineralogical data about the Mars surface. Drills on the rovers gained a beneath-the-dust view of the mineralogy of Mars.

Twenty years lapsed between the successful Viking spacecraft and the next Mars spacecraft success. In between, two Soviet Union probes were destroyed, and the American Mars Observer blew apart as it prepared to orbit Mars. Launched at the end of 1996, the Mars Global Surveyor (MGS) began its prime mapping mission in March 1999, after spending a year and a half gradually trimming its orbit. MGS set Mars mission records. In nine years of orbiting pole to pole, MGS sent back enough data to fill 1,000 CDs. It has mapped the entire planet. Its long service permitted monitoring of surface changes, including detecting recent impacts that were not present on earlier passes and detecting sediment patterns in craters that appeared to result from temporary liquid water seeps. MSG also monitored long-term Martian weather patterns and local effects such as dust devil tracks.

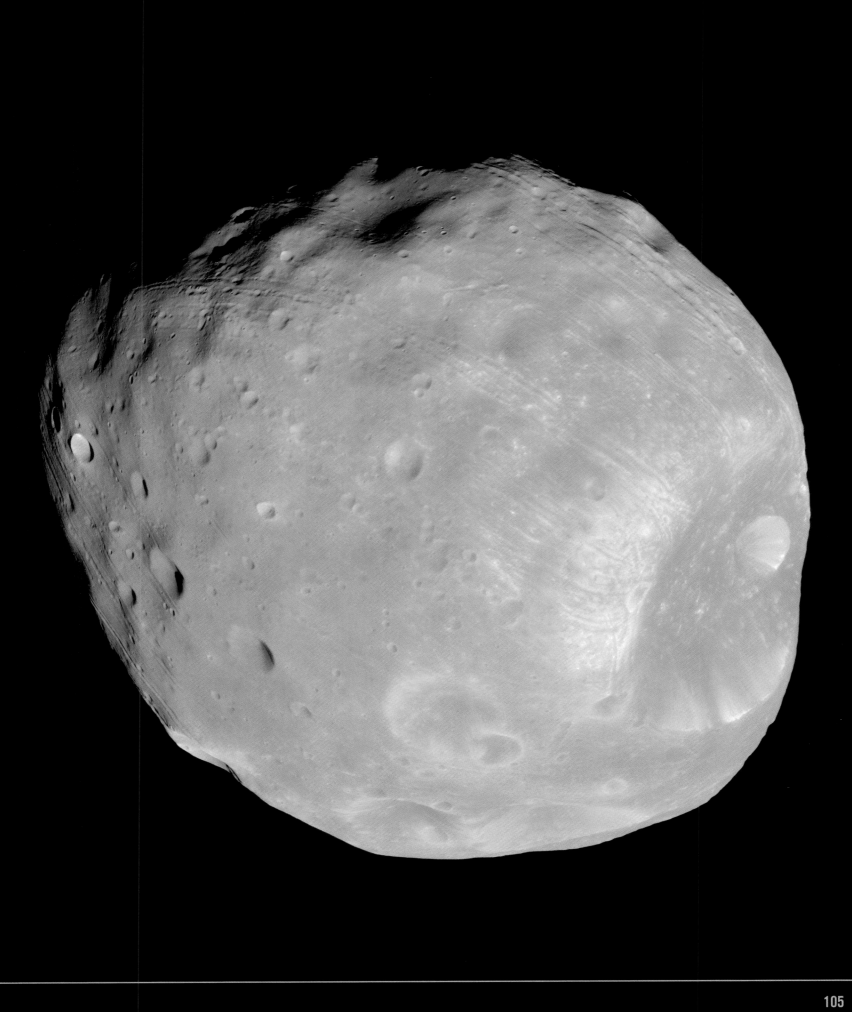

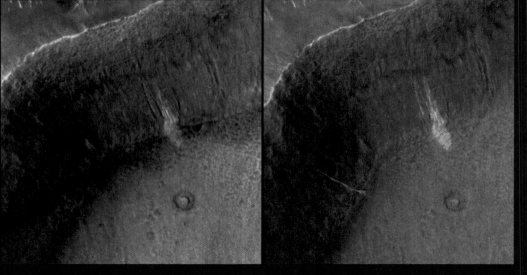

MGS's camera was used to test the hypothesis that some very youthful-looking gullies found on slopes at middle and high latitudes on Mars are so young that some of them could still be active today. The test was very simple: re-image gullies previously seen by the camera and see if anything had changed.

Do the images taken prove that water has flowed on Mars? No, but they strongly suggest it. Although Mars is drier than the most arid deserts on Earth, liquid water from beneath the Martian surface may have welled up and flowed across this portion of the Red Planet in this decade.

Image Credit: NASA/JPL/Malin Space Science Systems

The 2001 Mars Odyssey orbiter carried science experiments designed to make Mars-wide measurements in order to understand the planet's climate, geologic history, and radiation environment and to search for life-sustaining water. Ironically, the radiation instrument aboard Mars Odyssey was exposed to a solar storm of such magnitude that its electronics were "fried." The Martian radiation environment, along with the radiation environment experienced by spacecraft in transit from Earth to Mars, is of great concern to planners of future manned Mars missions. Unless a remedy is found, Mars astronauts may find themselves far too sick from radiation to function on the planet.

At the end of 2003, following a quick six-month trajectory, the European Space Agency Mars Express spacecraft began service. Its main objective was to search for subsurface water. Amazing images and other data are still being investigated, and hypotheses regarding their meaning are being debated. Unfortunately, a small lander craft, named Beagle 2, after Charles Darwin's Beagle, failed to land safely.

A Delta II rocket lofts the Phoenix spacecraft toward Mars on August 4, 2007. The mission's plan is to land in icy soils near the north polar permanent ice cap of Mars and explore the history of the water in these soils and any associated rocks, while monitoring polar climate. The spacecraft and its instruments are designed to analyze samples collected from up to a half-meter (20 inches) deep by a robotic arm.
Image Credit: NASA

The mission's main objective was to search for subsurface water from orbit. Spacecraft instruments focused on investigations designed to help answer questions about Mars's geology, atmosphere, and surface environment and look for signs of water and potential sites where life might be found. Among its discoveries are possible evidence of recent glacial activity, explosive volcanism, and methane gas.

The latest in the Mars spacecraft armada is the Mars Reconnaissance Orbiter, or MRO. Since beginning its primary science mission at the end of 1996, it has already exceeded the data record established by MGS. Among its discoveries are "haloes," markings left on underground rocks as liquids or gases have flowed through them. On Earth, such features often indicate ripe conditions for tough microbial life. MRO has also "reconned" the northern plains, where the future Phoenix Scout mission is planned to land in 2008. Knowing potential hazards before landing is extremely important to Phoenix planners.

With the string of recent spacecraft successes and interest in Mars as a target for future manned missions, the computer screens of mission planners are glowing with a new generation of spacecraft designs and space exploration tools. This new generation includes larger rovers, agile robots that can scale cliffs, microbots that randomly swarm from the lander mother ship, Mars airplanes, lighter-than-air balloons, and landers that will package Mars samples and rocket them back to Earth. If all goes well, humans will follow.

A mission carrying people to Mars will be daunting to plan. Travel to Mars is way above and beyond the challenges faced before the first Apollo flight in 1969. With chemical rockets, the Moon is three days away. Mars is three to six months away, depending upon how fast the vehicle can travel. Furthermore, Mars astronauts will have to remain on Mars for twelve months before Mars and Earth are in favorable positions for a return. The quantity of food, water, oxygen, and rocket fuel needed to support a two-year mission is far greater than the carrying capacities of any rocket being planned. Such a Mars mission will require a host of technologies

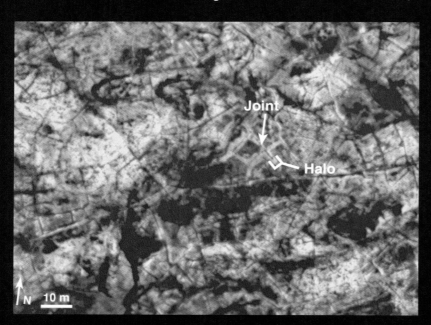

This image from the High Resolution Imaging Science Experiment camera on NASA's Mars Reconnaissance Orbiter shows evidence for ancient fluid flow along fractures in Mars' Meridiani Planum region. The scene includes pervasive signs of ancient fluid flow in the form of bleached and cemented features, called halos, along fractures within the layered deposits of Meridiani. This site is approximately 375 kilometers (233 miles) northeast of "Victoria Crater."
Image Credit: NASA/JPL/Univ. of Arizona

This engineering model of Mars Science Laboratory was dubbed "Scarecrow" by the mobility team, because it is still without a brain like the famous scarecrow from "The Wizard of Oz."

The 2009 Mars Science Laboratory, the mammoth grandchild of the 1997 Sojourner rover, is nearly assembled and ready for its test and launch operations phase (ATLO). With its immense increase in size come advanced abilities in power, technology, and science data collection.

Image Credit: NASA/JPL-Caltech

and scientific breakthroughs. Robotic greenhouses and oxygen, water, and rocket fuel processing plants for returning to Earth will have to be sent in advance of a crew. Only when the infrastructure of a Mars base is up and running will people risk the trip. But what an adventure it will be! Perhaps future Mars explorers will discover new forms of life. Then, for the first time, we will know for sure that Earth is not the only oasis of life in the universe.

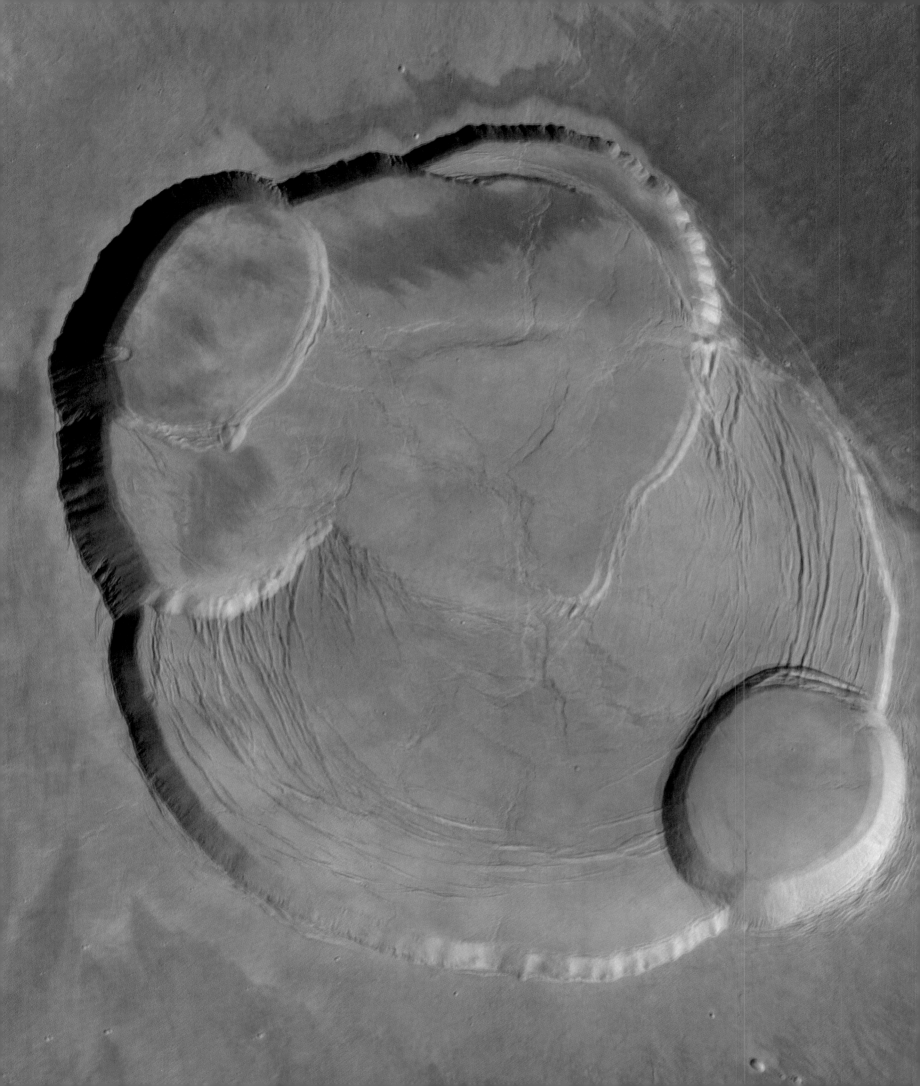

10 MARS IN 3D

Two eyes are needed to see the world in three dimensions. Each eye sees from a slightly different angle. Our brains process the images in a way that permits us to perceive depth.

When we look at Mars through even the biggest Earth telescopes, Mars is two-dimensional, a flat disk. The through-the-telescope view is the same kind of view we get when we close one eye. Our vision becomes flat. Fortunately, Mars spacecraft provide enough image data to bring back the third dimension. It's a simple trick. A "one-eye" camera can provide the 3D view by taking two pictures of the same scene but from slightly different positions. This mimics looking at the same scene with two eyes. You can perform this trick with a digital camera. Take a picture and then shift the camera to the right or left about 8 cm (3 in.) Take a second picture.

The next step is to load your pictures onto your computer and process them with an image program such as PhotoShop. Check the reference section for an Internet site with directions on how to do this.

With the following Mars pictures, two nearly identical but slightly shifted scenes are overlaid. Each image is given a color tint, usually one red and one blue. In normal vision, the picture is blurry. However, with 3D glasses, a red lens to filter out the blue image and a blue lens to filter out the red image, the two-eye effect takes place. Your brain combines the two images and—abracadabra: a three-dimensional picture of Mars!

Use the 3D glasses provided with this book to look at the pictures in this chapter. You can easily make additional glasses out of stiff cardstock paper and red and blue acetate.(Acetate is available from art and craft supply shops.) Cut out a pair of cardboard frame glasses to fit your eyes. Tape a rectangle of red acetate over the left eyehole and a piece of blue acetate over the right eye hole. If you like, decorate the frames to make a personal fashion statement. Otherwise, the 3D glasses are finished!

This image was taken by the stereo cameras on ESA's Mars Express. It shows the western flank of the shield volcano Olympus Mons in the Tharsis region of Mars's western hemisphere. The escarpment at the lower left rises to over 7,000 meters. At the top of the image, part of the extensive plains west of the escarpment can be seen; these plains are known as the "aureole" (from the Latin for 'circle of light'). This region contains gigantic ridges and blocks extending some 1,000 kilometers from the summit, like the petals of a flower. The origin of the deposits in this region is still a mystery. The most persistent explanation, however, has been landslides. Several indications also suggest a development and resurfacing connected to glacial activity.
Image credit: ESA/DLR/FU Berlin (G. Neukum)

One of the very first places photographed by the Mars Orbiter Camera (MOC) as part of the MGS mission at the start of its mapping phase in March 1999 was a field of dunes located in Nili Patera, a volcanic depression in central Syrtis Major. On April 24, 2001, MGS was pointed off-nadir to take a new picture of the same dune field. By combining the views from March 1999 and April 2001, a stereoscopic image was created, the anaglyph shown here. The dunes and the local topography of the volcanic crater's floor stand out in sharp relief. The images, taken more than one Mars year apart, show no change in the shape or location of the dunes.
Image credit: NASA/JPL/Malin Space Science Systems

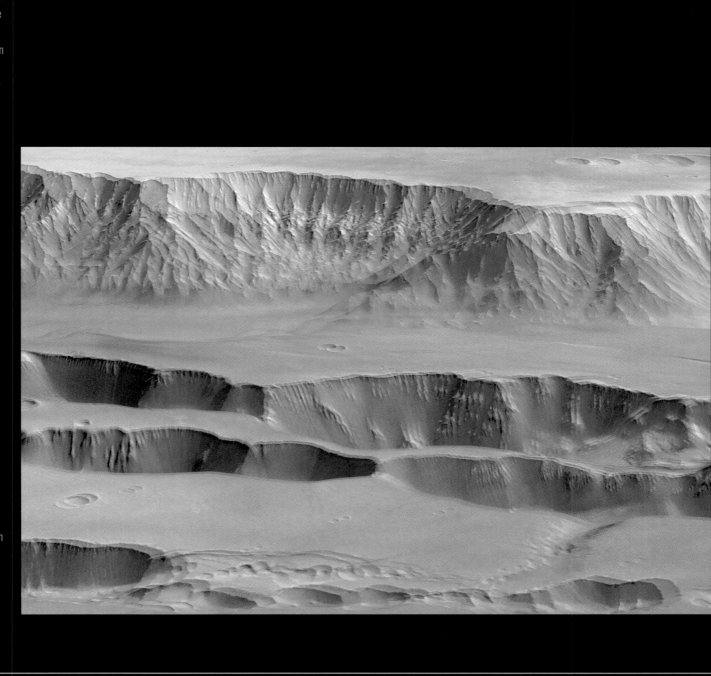

This anaglyph uses two MOC images acquired at slightly different viewing angles. Owing to the specifics of the viewing geometry, the image is tilted on its side. In other words, in this image, north is toward the right and west is up. In this picture you can see better the cross-cutting between layers in the mound located in southern Galle (Happy Face) Crater. The layers are part of a mound of sedimentary rock in southern Galle, a remnant of a once more extensive deposit of sedimentary material in this south mid-latitude impact basin.
Image credit: NASA/JPL/Malin Space Science Systems

The HRSC on ESA's Mars Express obtained this 3D perspective view in May 2004. The scene shows the region of Coprates Chasma and Coprates Catena.
Image credit: ESA/DLR/FU Berlin (G. Neukum)

Yogi is a meter-size rock a short distance from the Mars Pathfinder lander and was the second rock visited by the rover Sojourner's alpha proton X-ray spectrometer (APXS) instrument. This view was produced combining the "Super Panorama" frames from the IMP camera. Super resolution was applied to help to address questions about the texture of this rock and what it might tell us about its mode of origin.

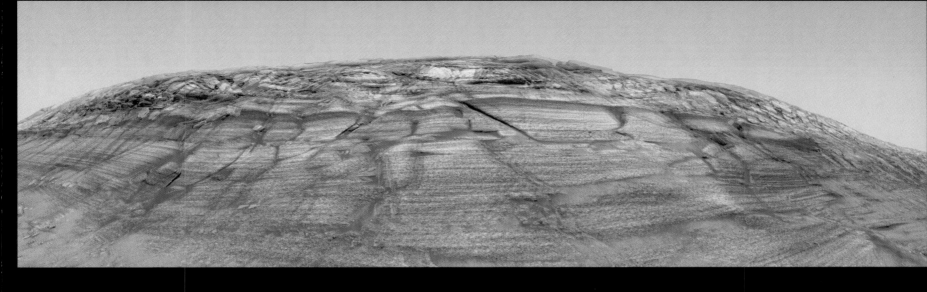

The panoramic camera on NASA's rover Opportunity captured a sweeping image of "Burns Cliff" after driving right to the base of this southeastern portion of the inner wall of "Endurance Crater" in November 2004. Because of the wide-angle view, the cliff walls appear to bulge out toward the camera. In reality the walls form a gently curving, continuous surface.

Image credit: NASA/JPL/Cornell

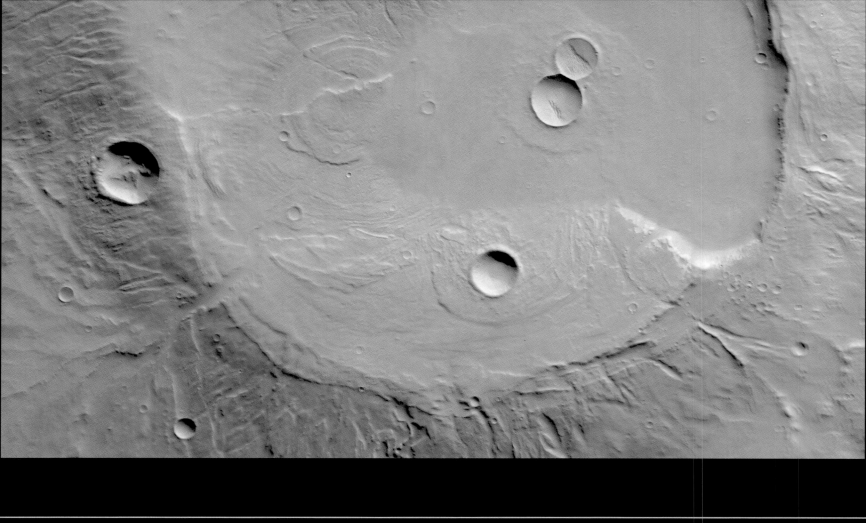

The Mars rover Spirit maneuvered along the edge of an arc-shaped feature called "Lorre Ridge" and encountered some spectacular examples of basaltic rocks with striking textures. This panoramic image shows a group of boulders informally named "FuYi." The rocks were formed by volcanic processes and may be a primary constituent of Lorre Ridge and other interesting landforms in the basin. Spirit first encountered basalts on a vast plain covered with solidified lava that appeared to have flowed across Gusev Crater. The basaltic rocks at Lorre Ridge exhibited many small holes or vesicles, similar to some kinds of volcanic rocks on Earth. Vesicular rocks form when gas bubbles are trapped in lava flows and the rock solidifies around the bubbles. When the gas escapes, it leaves holes in the rock. The quantity of gas bubbles in rocks varied considerably; some rocks had none and some, such as several here at FuYi, were downright frothy.
Image credit: NASA/JPL-Caltech/Cornell

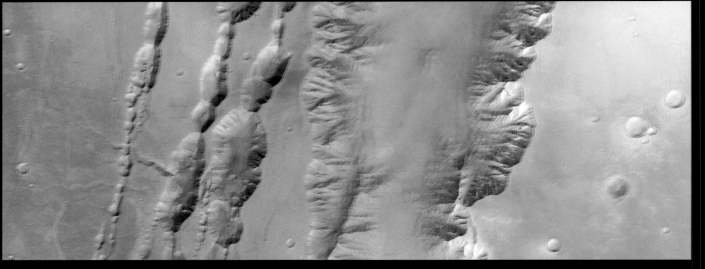

This anaglyph image was taken by the HRSC on board ESA's Mars Express spacecraft. It show Coprates Chasma, a major trough located roughly in the center of the Valles Marineris canyon system.
Image credit: ESA/DLR/FU Berlin (G. Neukum)

At the center of this picture, Sojourner has traveled off the lander's rear ramp and onto the surface of Mars. The rock Barnacle Bill is to the left of Sojourner, and the large rock Yogi is at upper right. On the horizon sits the rock dubbed "Couch." A deflated airbag sits at lower right.
Image credit: NASA/JPL

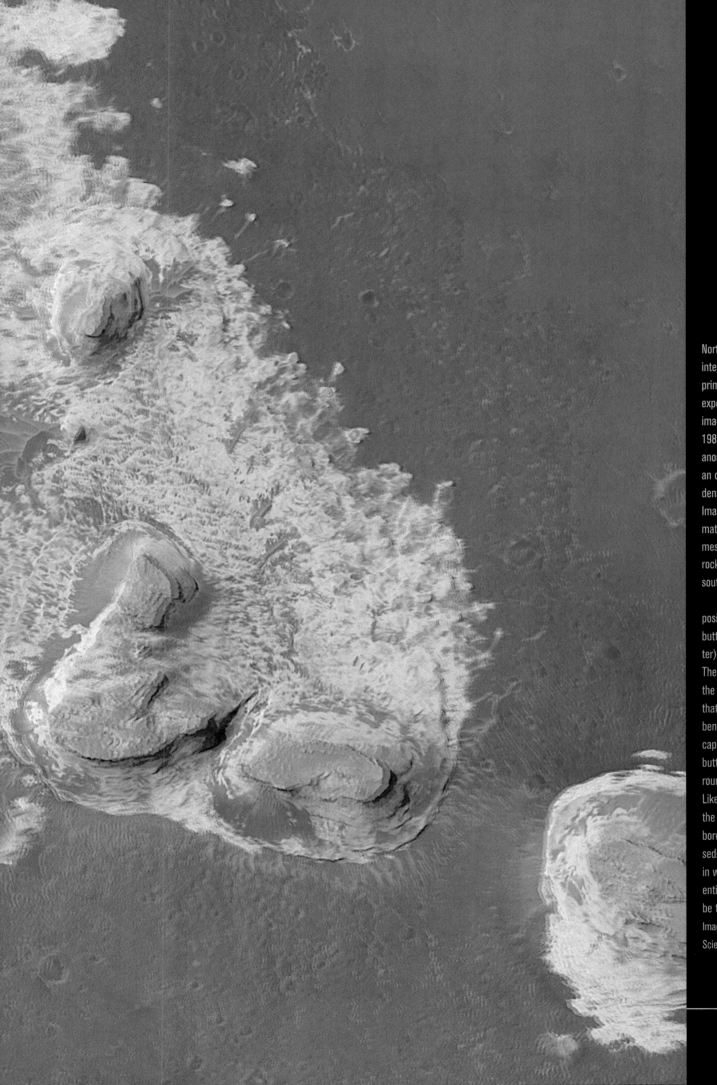

Northern Terra Meridiani, near the intersection of the Martian equator and prime meridian, is a region of vast exposures of layered rock. A thermal image from the Phobos 2 orbiter in 1989 showed these materials to be anomalously cool during the daytime, an observation very suggestive of dense, hardened materials like rock. Images of this region showed layered material exposed in cliffs, buttes, and mesas that in some ways resemble the rock outcrops of northern Arizona and southeastern Utah in North America.

The image shown here is a composite of two pictures. It shows four buttes and a pinnacle (near left-center) composed of eroded, layered rock. The four buttes are each capped by the remains of a single layer of rock that is harder than the materials beneath it. It is the presence of this caprock that has permitted these buttes to remain standing after surrounding materials were eroded away. Like the buttes of Monument Valley in the Navajo Nation on the Arizona/Utah border, these are believed to consist of sedimentary rocks, perhaps deposited in water or by wind, though some scientists have speculated that they could be thick accumulations of volcanic ash.
Image credit: NASA/JPL/Malin Space Science Systems

The High Resolution Imaging Science Experiment on NASA's Mars Reconnaissance Orbiter has imaged "Victoria Crater" three times. This stereo view combines two of those views. The difference in viewing angle between the two images is about 12 degrees, which is greater than the convergence angle between the left and right eyes of humans while viewing distant objects, so the vertical relief appears much steeper than is actually the case. Although some of the cliffs around the crater are, in fact, vertical, the slopes below the cliffs are no steeper than 30 degrees.
Image credit: NASA/JPL-Caltech/University of Arizona

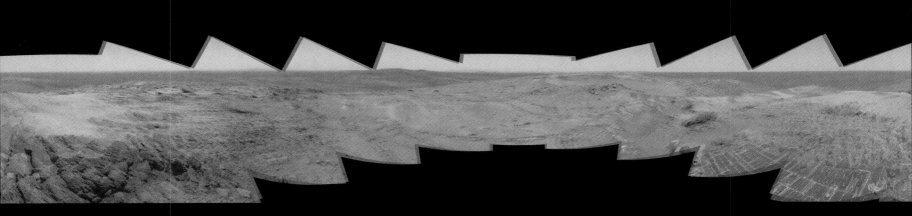

Before moving on to explore more of Mars, the rover Spirit looked back at the long and winding trail of twin wheel tracks it created to get to the top of Husband Hill. Spirit spent several days perched on a lofty, rock-strewn incline next to a precarious outcrop nicknamed "Hillary." Researchers helped the rover make several wheel adjustments to get solid footing before conducting scientific analysis of the rock outcrop. To the west are the slopes of the "Columbia Hills," so named for the doomed space shuttle Columbia. Beyond the hills are the flat plains and rim of Gusev Crater.
Image credit: NASA/JPL-Caltech

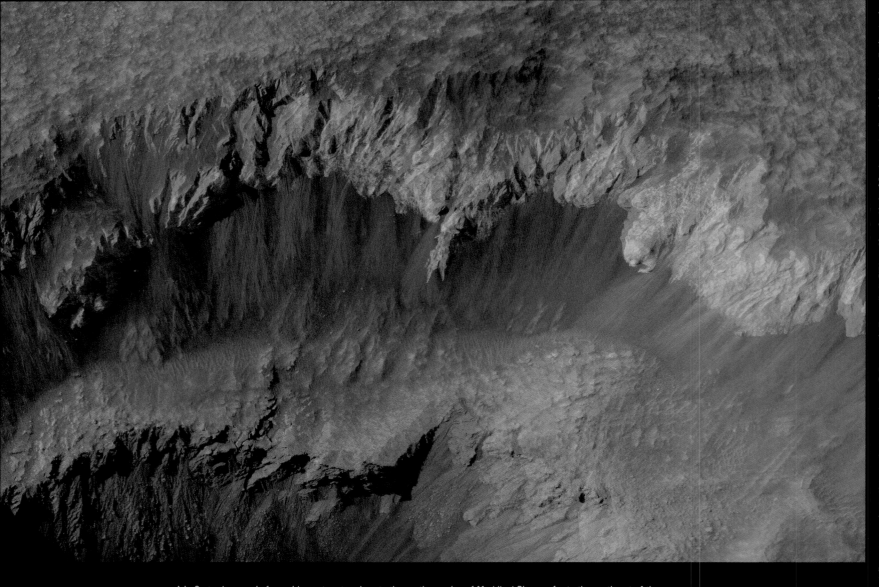

Ada Crater is a newly formed impact crater close to the southern edge of Meridiani Planum, far to the southeast of the Opportunity rover. The crater looks extremely deep, but that impression is greatly exaggerated! Stereo pairs are acquired with separation angles much greater than that of our own eyes, in order to extract for accurate measurements. But the effect on color anaglyphs is to exaggerate the relief, which can be vertigo-inducing over steep terrain.

Image credit: NASA/JPL/University of Arizona

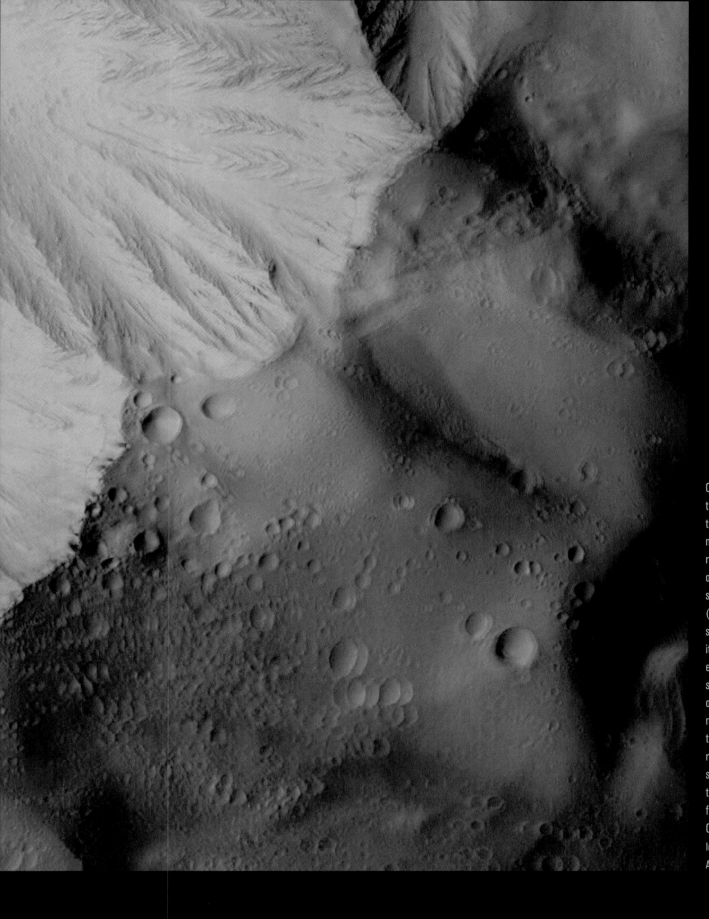

Olympus Mons is the largest volcano in the Solar System. Although it is known that it was constructed of lava flows, many aspects of this titanic volcano remain puzzling. For example, the base of the volcano is marked by a steep scarp (cliff) that is up to 8,000 meters (26,000 feet) tall. The volcano may be so large that it is falling apart under its own weight, such that the outer edges are collapsing in massive land-slides. By combining two HiRISE observations, we can see an approximately 6 km (3.7 mile) wide portion of this scarp in three dimensions. The rugged topography at the edge of the scarp, with kilometer-scale pieces of the volcano pushed up or pulled apart, fits the idea that the lower part of Olympus Mons is riddled with faults.
Image credit: NASA/JPL/University of Arizona

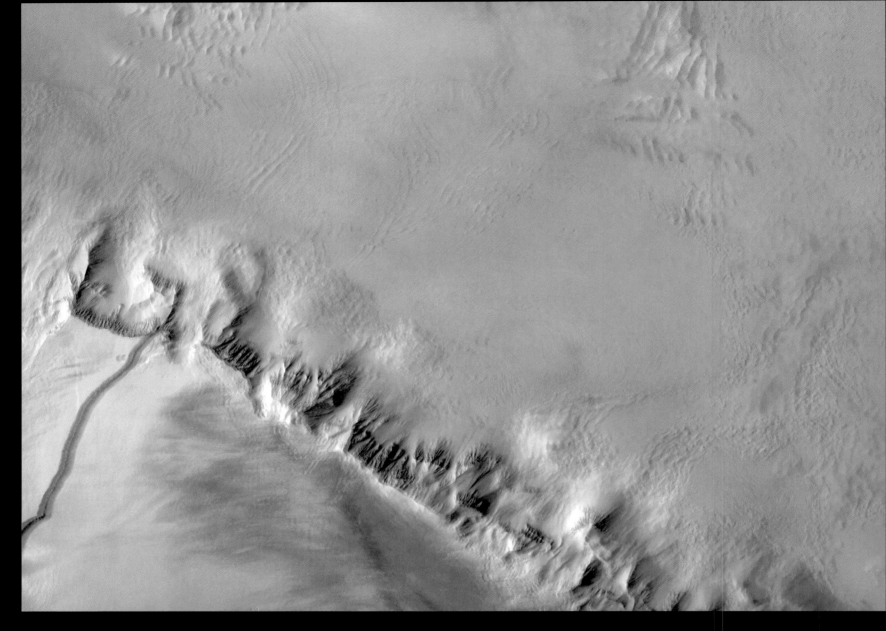

This image shows the western flank of the shield volcano Olympus Mons in the Tharsis region of the Martian western hemisphere. The escarpment at lower left rises from the surface level to over 7,000 meters. At the top of the image, part of the extensive plains west of the escarpment are shown.

To the northwest of the volcano are regions of gigantic ridges and blocks extending some 1,000 kilometers from the summit like the petals of a flower. The origin of the deposits has challenged planetary scientists for an explanation for decades. The most persistent explanation, however, has been landslides.

Image credit: ESA/DLR/FU Berlin (G. Neukum)

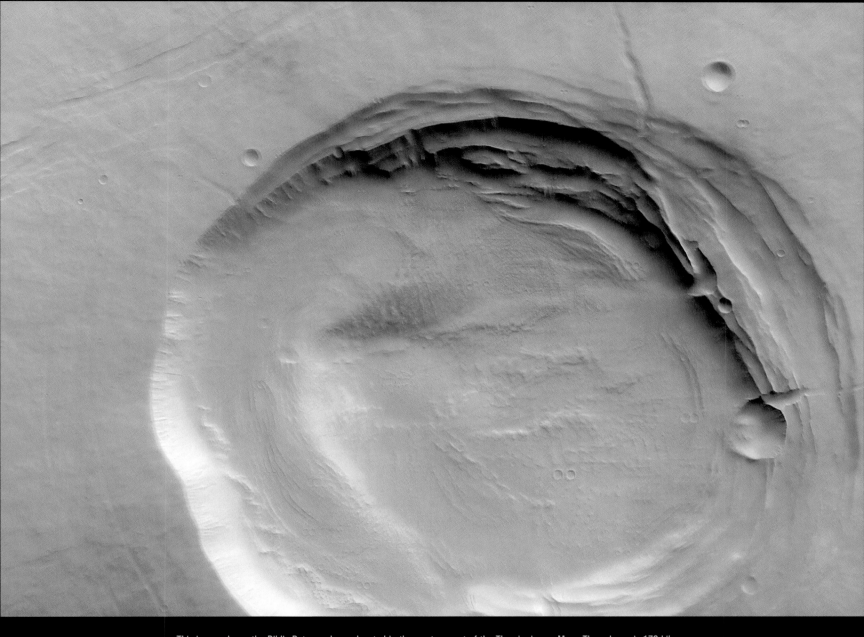

This image shows the Biblis Patera volcano, located in the western part of the Tharsis rise on Mars. The volcano is 170 kilometers long, 100 kilometers wide, and rises nearly 3 kilometers above its surroundings. The bowl-shaped 'caldera' has a diameter of 53 kilometers and extends to a maximum depth of roughly 4.5 kilometers.

Image credit: ESA/DLR/FU Berlin (G. Neukum)

In early spring in the southern hemisphere on Mars the ground is covered with a layer of carbon dioxide ice. In this image there are two lanes of undisturbed ice bordered by two lanes peppered with fans of dark dust. When we zoom in to the sub-image, the fans are seen to be pointed in the same direction, dust carried along by wind. The fans seem to emanate from spider-like features. The arms are channels carved in the surface, blanketed by the carbon dioxide ice. The ice, warmed from below, evaporates, and the gas is carried along the channels. Wherever a weak spot is found the gas vents to the top of the ice, carrying along dust from below. These channels are deepening and widening as they converge. Spiders like this are often draped over the local topography, and often channels get larger as they go uphill. A different channel morphology is apparent in the lanes not showing fans. In these regions the channels are more like lace and are not radially organized.

Image credit: NASA/JPL/University of Arizona

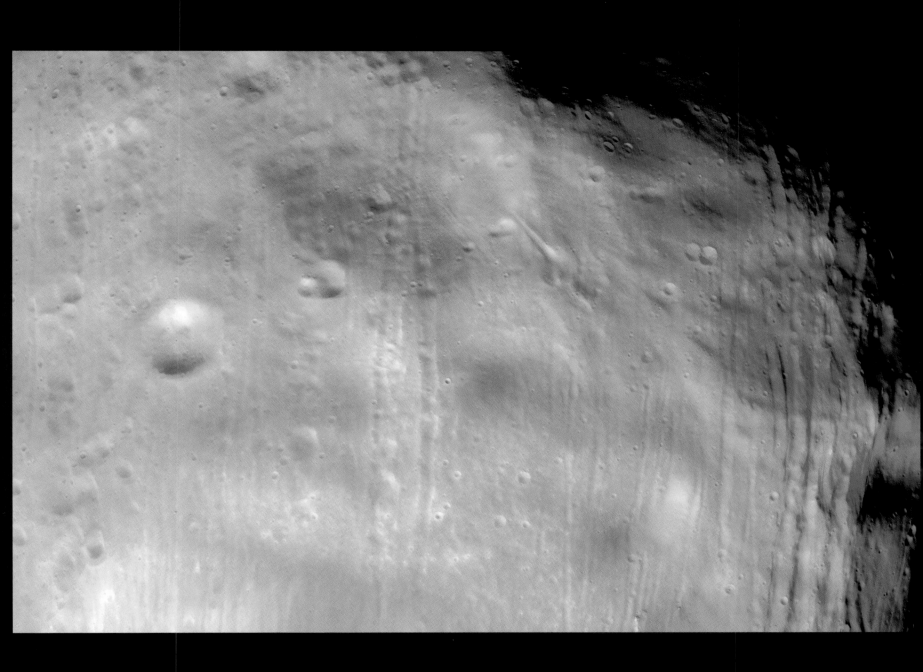

Phobos and Deimos are interesting for several reasons. Both Martian moons are small, with average diameters of just 22 and 12 km, respectively. At this size, their gravity is insufficient (less than 1/1000 of Earth) to pull them into spherical shapes, in contrast to the larger moons and planets in the Solar System. Both moons are tidally locked toMars, meaning, like our own Moon relative to Earth, they present the same side to Mars all the time. The small size and composition of Phobos and Deimos make them very similar to some asteroids. It is possible, in fact, that they are captured asteroids. Other hypotheses are that they formed with Mars in the early Solar System, or are composed of material blasted off of Mars by impacts.

Image credit: NASA/JPL/University of Arizona

128

APPENDICES

APPENDIX A

TOPOGRAPHIC TERMINOLOGY

As spacecraft made one startling discovery after another, Mars scientists decided that Mars needed its own nomenclature. They elected to employ Latin terms for topographic features and consequently, some explanation is usually necessary.

chasma (pl. chasmata)	steep-sided elongated depression
fossa (pl. fossae)	long and narrow valley
labyrinthus (pl. labyrinthi)	interconnecting valleys
mons (pl. montes)	mountain
patera	shallow crater, usually volcanic in origin
planitia (pl. planitiae)	low plain
planum (pl. plana)	plateau
rima	narrow fissure
rupe	scarp or cliff
terra (pl. terrae)	extensive landmass region
tholus	small domed volcano
vallis (pl. valles)	valley
vasitas	wide lowland

MARS VS. EARTH: THE FACTS

Feature	Mars	Earth
Orbit rank (from the Sun)	4th	3rd
Orbit (semi-major axis 106 km)	227.92	149.60
Perihelion (106 km)	206.62	147.09
Aphelion (106 km)	249.23	152.1
Average orbital velocity (km/s)	24.13	29.78
Orbit inclination (in degrees)	1.85	0.0
Orbit eccentricity (flattening)	0.0935	0.0167
Orbit period (Earth days)	686.980	365.256
Rotation (Earth hours)	24.6597	24.000
Equatorial radius (km3)	3397	6378.3
Polar radius (km3)	3375	6356.8
Mass (1024 kg)	0.64185	5.9736
Mean density (kg/m3)	3933	5515
Surface gravity (m/s2)	3.71	9.8
Escape velocity (km/s)	5.03	11.19
Topographic range (km)	30	20
Atmosphere	95% CO_2, remainder N, Ar, O_2, CO, H_2O, NO, Ne	78.084% N_2, 20.946% O_2, remainder Ar, CO_2, Ne, He, CH_4, K, H_2
Surface pressure (millibars)	4.0 to 8.7	1014
Temperature range	~ 210K (?63° C)	288K (15° C)
Natural satellites	2	1

ROBOT MARTIAN EXPLORERS *

Year	Mission	Origin	Status
1960	*Marsnik 1*	USSR	Launch Failure
	Marsnik 2	USSR	Launch Failure
1962	*Sputnik 22*	USSR	
	Mars 1	USSR	Contact lost in transit
	Sputnik 24	USSR	Mars transit trajectory failure
1964	*Mariner 3*	USA	Launch failure
	Mariner 4	USA	Flyby
	Zond 2	USSR	Partial power failure, contact lost
1965	*Zond 3*	USSR	Mars launch window missed, mission retasked as spacecraft test
1969	***Mariner 6***	USA	Flyby
	Mariner 7	USA	Flyby
	Mars 1969A	USSR	Launch failure
	Mars 1969B	USSR	Launch failure
1971	*Mariner 8*	USA	Launch failure
	Cosmos 419	USSR	Mars transit trajectory failure
	Mars 2	USSR	Orbited Mars, lander crashed
	Mars 3	USSR	Orbited Mars, soft landing – lander failure after 20s
	Mariner 9	USA	Orbited Mars

Year	Mission	Origin	Status
1973	Mars 4	USSR	Failed to orbit, became flyby mission, some data returned
	Mars 5	USSR	Orbited Mars
	Mars 6	USSR	Contact lost during landing
	Mars 7	USSR	Failed to orbit, lander missed Mars, both went into heliocentric orbit
1975	Viking 1	USA	Orbited and Landed
	Viking 2	USA	Orbited and Landed
1988	Phobos 1	USSR	Power failure, contact lost in transit
	Phobos 2	USSR	Contact lost
1992	Mars Observer	USA	Contact lost
1996	**Mars Global**		
	Surveyor	USA	Orbited Mars
	Mars 96	USSR	Launch failure
	Mars Pathfinder	USA	Rover on Mars
1998	Nozomi (Planet B)	Japan	Failed to orbit, went into heliocentric orbit
	Mars Climate Orbiter	USA	Failed to orbit, crashed on Mars
1999	Mars Polar Lander	USA	Contact lost on atmosphere entry
	Deep Space 2	USA	Contact with both probes lost
2001	Mars Odyssey	USA	Orbited Mars
2003	Mars Express	European Space Agency	Orbited Mars, lander lost
	Spirit	USA	Rover on Mars
	Opportunity	USA	Rover on Mars
2005	**Mars Reconnaisance Orbiter**	USA	Orbited Mars
2007	Phoenix	USA	Planned lander
2009	Mars 2009	USA	Planned laboratory rover
2011	Mars 2011	USA	Mission TBD
2014+	Not selected yet	USA	Planned sample return

▪ Mission names in bold were fully or partially successful.

MARS OPPOSITIONS: WHEN THE VIEW OF MARS IS BEST

An opposition of Mars occurs when Mars and the Sun are on opposite sides of Earth. This is the best time to observe Mars because it is when it is closest to Earth.

The interval of oppositions normally ranges between 25 and 28 months, due to the different orbital periods of the two planets, greater eccentricity (flattening) of Mars's orbit, and the gravitational tugs of Jupiter and other planets. Mars takes 1.881 Earth years to orbit the Sun. During its orbit, the planet ranges from 207 to 249 million km (121 to 155 million mi) from the Sun. Earth's orbit is nearly circular and only varies from 147 to 152 million km (91 to 94 million mi). Earth's greater orbital velocity (about 30 km or 18.6 m per second) causes it to swing past the slower Mars (24 km, or 14.9 mi per second) and race around the Sun to catch up with Mars again about 26 months later.

The Mars Opposition of 2003

On August 28, 2003, Mars and Earth were in opposition, and Mars was only 55.76 million km (34 million m) away. The disk of Mars spanned 25.11 arc seconds (about one-half of one-sixtieth of a degree). For comparison, the Moon spans about one half of a degree (1,800 arc seconds). Although not sounding like much, the opposition was about as good as it gets for Mars. The last time it was that close, Neanderthals were roaming Europe. The next good opposition of Mars will be on July 27, 2008, with Mars coming within 58 million km (35.38 million m) of Earth and presenting a disk 24.31 arc seconds across.

Opposition Calendar*	Apparent Diameter of Mars (arc seconds)	Distance (millions of km)
2010 January 29	14.1	99.33
2012 March 3	13.89	100.78
2014 April 8	15.16	92.39
2016 May 22	18.6	75.28
2018 July 27	24.31	57.59
2020 October 13	22.56	62.07
2022 December 8	17.19	81.45
2025 January 16	14.57	96.08
2027 February 19	13.81	101.42
2029 March 25	14.46	96.82
2031 May 4	16.91	82.78
2033 June 27	22.13	63.28
2035 September 15	24.61	56.91
2037 November 19	18.96	73.84
2040 January 2	15.32	91.39
2042 February 6	13.93	100.49
2044 March 11	14.03	99.79
2046 April 17	15.68	89.32
2048 June 3	19.76	70.86
2050 August 14	25.02	55.96
2052 October 8	21.23	65.96
2054 December 17	16.42	85.29
2057 January 24	14.28	98.06
2059 February 27	13.83	101.25

* http://www.seds.org/~spider/spider/Mars/marsopps.html

Printed in the United States of America